"We may not know until the last heartbeat of the universe what effect one single choice has upon our lives; but it is true that the sum of all our choices will determine our exaltation or damnation."

THE LIGHTNING AND THE STORM

A NOVEL

by

MARSHA NEWMAN

WELLSPRING PUBLISHING
Salt Lake City, Utah

With the exception of actual historical personages identified as such, the characters of this book are entirely the product of the author's imagination.

Copyright © 1986 by Wellspring Publishing & Distributing
All rights reserved
Printed in the United States of America
First printing April 1986
Second printing July 1989

ISBN 0-9608658-2-9

Lithographed in the United States of America
PUBLISHERS PRESS
Salt Lake City, Utah

Written in honor of all those who struggle with the challenges and choices of life.

ACKNOWLEDGEMENTS

We express our appreciation for all those who contributed to the writing and editing of *THE LIGHTNING AND THE STORM* . Our thanks go to Judy Bendorf, not only for encouragement but also for information and background on horses, and to Marjorie Larsen for reading the manuscript and encouraging us to publish it. We are also appreciative of those who helped in the editing and proofing of the manuscript, particularly Darla Hanks, Sheila Woods, Harriet Woolley and Janis Bailey. For the cover photo we are most indebted to Kathie Horman for her time in helping us find the models and location.

CHAPTER 1

She wasn't the slightest bit repentant. Charlotte O'Neill was hardly ever repentant. She had that perfect assurance of nature that told her that whatever she wanted must be right, and she rarely bothered to question that assurance. Today she had hurried impatiently through the household chores, the mending of her brother's trousers, and the shaping of three loaves of bread. While her mother was engaged in correcting Annie's needlepoint, she had slipped out of the back door, unhitched their mare from her slow path around the flour mill, and ridden off bareback through the thicket behind the house. She knew where the trail would bring her out—into the meadow about two miles away. It was her favorite place. In the spring the grass was freckled with wild flowers: bluebells, baby's breath, sweet william and morning glory. Close to the long line of trees that gently cradled the meadow and by the banks of the shallow stream, grew masses of blackberries. She always came home with purple lips and slightly blue fingers.

The grass grew thick and green, right down to the water's edge. She had turned Esther loose to munch the grass and flowers, and she had gone wading in the stream, skirts hiked up above her knees, pantaloons clinging damply to her legs. If asked, she would have admitted that her father was going to be very unhappy that she had taken Esther from the mill that ground their flour. But there was no one to ask her, and she never asked herself uncomfortable questions. When the flour would be ground was the furthest thing from her mind. What her mother would say to her slipping away from her embroidery lessons mattered little when she was basking in the May sunlight, toes pink from the chill water of the stream.

Charlotte was next to the youngest in a family of three children. There were two girls and one boy living. Her mother had borne five children in all, but two boys had died at birth. Now, with John Patrick on a mission in the East, the rough board house contained Charlotte, barely seventeen, Annie, her quiet younger sister, Margaret and Patrick O'Neill.

Charlotte's warm brown hair came from her mother, Margaret, and the abundant threads of polished copper from Patrick's Irish head. He had inherited a luxurious thatch of red hair from his parents who had immigrated to America when he was fifteen. Similarly, he had inherited a short-fused Irish temper and passed it on to his young daughter. Sometimes there was fire and lightning between them, but sometimes there was an adoring love such as only passionate souls can have. Patrick had known since she came into the world, kicking and bawling at the top of her voice, that this little girl was his own kindred spirit. It was Charlotte that he thought to bring treats for, and Charlotte that he danced around the barn on his shoulders. It was her adoring hugs and kisses that brought unashamed tears to his eyes. And it was her willful ways and stamping foot of defiance that popped the cork of his temper. It was always a cuff and a hug. He would smack her coppery head in quick anger and immediately grab her and kiss her forehead in heart-broken regret. Patrick fortunately possessed a certain humility that allowed him quick repentance from the rashness of his anger. But Charlotte's pride was such a force that it petrified her angry words, and they were recalled only with great effort.

Normally her eyes were gray-green, subtly dotted with golden flecks, but excitement or anger turned them emerald green. They were a sure barometer of her mood.

"What're ye green about now, my princess?" Patrick would ask, while her mother commented in her tight-lipped way, "Green's too flashy a color for a lady. Better mind your manners."

Her cheek bones were high and wide, her nose delicate and slightly patrician, and her mouth full and generous. She was a little taller than other girls her age, but slender as a willow wand. There were times Margaret watched her daughter bent over chores in the house and she thought her rather plain. Then Charlotte would glance up, her eyes filled with curiosity, or excitement, and her mother glimpsed the spark that transformed her with a beauty all her own.

The whole O'Neill family had been baptized on a chilly March day at Far West, Missouri in 1838—the whole family except Charlotte and

Annie. Annie was too young, as the doctrine of the Church was to baptize only persons eight years of age or older. Charlotte was twelve and already possessed of a mind of her own, which, on this subject was very made up. She would not be baptized just because every one else was. She had heard the story of the Book of Mormon and the stories of Joseph's prophecies. After a period of skepticism, Patrick O'Neill became convinced that young Joseph Smith was really a prophet. Several long nights were spent in their little farm outside Far West, Missouri, discussing the Church, and finally the whole family agreed that it was true— everyone except Charlotte. She would say nothing. They had prayer together each night, kneeling on the cold, rough-boarded floor. One by one, her sister and brother, father and mother came to life with the spirit of the Mormon gospel, and, one by one, they bore their newly found testimonies to each other. Charlotte was the only one who remained silent.

She usually did the milking and the lighter chores around the barn, and often Patrick would come in to help her, especially on cold mornings. He could hardly bear to see her come in the house with her hands all red and chapped. Many was the time he wished he had had other boys besides Johnny to help with the outside chores, so that his girls could do only women's work. One day while they were in the barn together, Charlotte milking Daisy and Patrick pitching hay into the upper loft, he brought up the subject of the baptism.

"Ye are quiet lately, my pet. Is there anything a holdin' onto yere tongue?"

"No, Papa," she answered him, knowing what was coming.

"Ye know we're all being baptized next Sabbath day. Are ye coming with us, into the water, I mean?"

"I can't, Pa."

"Why not, lassie?"

"I don't know why. I just don't want to do it."

"What stays ye?"

"I just don't know."

He waved his hand impatiently in the air. "Ah, ye are just being stubborn as ye're ever wont to be."

She pressed her lips more tightly together and concentrated on the milking.

"Ye'll be sorry, lass. Ye'll be sorry. It's all the truth, and God knows it. Ye know it, too, or I miss my guess." He waited a minute, watching her intently, struggling to know what to say to bring her around to his way of thinking. Nothing came to him. He shoved his big, thick hand through his copper-gold hair and stood up abruptly.

"Charlotte, lass, enough of this foolish stubbornness of yours. I want ye to be ready to go down into the waters next Sabbath day with the rest of us. Let yere good sense run ye this time instead o' yere will."

She shook her head minutely, but he saw the movement, and his short supply of patience ran out like grain upon the ground. "Ye will!" he shouted. "Ye will or ye can leave me house! I want no child such as ye acting the mule. All yere life have ye been thus, following no one's will except yere own. It's time ye learned to mind yere elders."

She caught his fire and jumped up, upsetting the bucket of milk to run all over his feet, lost in the filthy straw of the barn. In anger she lapsed into the same Irish brogue she had heard since a baby. "I won't and ye can na' make me! If ye want to follow like a bleating sheep after Joseph Smith, ye can baa all ye like." She stamped her foot and returned him glare for glare, as much like him as the sapling is the tree. "But he does na' convince me, and ye do na' convince me, and I will na' be baptized to something I do na' want to follow."

He roared at her then, and she stumbled backwards catching herself as she fell, then jumped up and dashed out as he advanced on her. She took to the fields, walking straight to the trees by the meadow, staying away for the rest of the day. As twilight overtook her, she began to wonder if her father really meant his threat about her leaving. Where would she go? The prospects were dismal, but she never cried out there alone in the fields facing banishment at twelve years old. There was a streak of iron in all the passion of her young soul. That iron, combined with the fiery temper that Patrick had passed on to his daughter, was a dangerous combination. If only he had not been so quick to anger himself, he might have been able to persist in loving past the point of her stubbornness, for that point did exist. In the face of continued love and gentleness, Charlotte would eventually bend. Only Annie and John Patrick, now so far away, had the perseverance to find that point.

From the time Annie was a baby and Charlotte a child of five, they had taken care of each other. As was the custom, the next oldest child took responsibility for the baby. Charlotte did almost everything for Annie. At first, when the baby was weeks old, Charlotte would sit by the hour rocking her, and as Annie grew older, she was the one who taught her to talk and coached her first staggering steps. Eventually they became playmates, as unlike as sugar and salt. Charlotte was full of imagination and reckless daring. Annie followed cautiously after, tugging on her sister's skirt whenever her tongue threatened to get her into trouble. Annie was a gentle soul, born to daydreams and protection. Her light brown hair, like Margaret's, showed few traces of the copper that brightened Charlotte's mane. Never once in her young life had there been a need to whip her, and she cried into the pillows of the bed whenever she heard the whistling switches on her sister's legs.

"If you'd only say you're sorry, or at least cry when he whips you, he wouldn't whip you so hard. It's just that he gets so mad when you don't cry like anyone ought to," Annie advised Charlotte.

But Charlotte would only say the more vehemently, "Let him be mad. I'll not cry like a puppy, and slink around with my belly to the ground just to keep from a switching."

Between the two tempers, Margaret was the restraining influence. The house was her dominion, and she had the last word there. Consequently, the few times Patrick took the switch to Charlotte inside the house, a word from Margaret told him it was enough. If he tried to ignore her, she came and took his arm, her own grey eyes beginning to turn to rock. "Tis enough, Patrick," she would warn him, her voice hard as her jaw.

If she saved her daughter from many a well-deserved switching, Charlotte had a hard time being thankful. For, though she rarely aroused her mother's ire, neither could she wheedle favors from her as she could Pat. Margaret was as cool and impartial as a judge, and the moments of real emotion between them were few. Once Esther's baby colt caught the animal sickness that was going around the community and had to be shot. The afternoon that the shot rang out, Charlotte jumped as though she herself had been the mark. Margaret dropped the spoon into the soup she was stirring and gathered her daughter up into her arms. They slid down to the floor and sat there for many long minutes, rocking and crying—Charlotte for the animal, Margaret for her half-grown daughter.

"Charlotte, Papa's going to have your hide! Esther was grinding the flour for the Stout family, and it was supposed to be ready before evening today." Annie had been sent to find her truant sister, and she knew just where to look.

"Look, Annie. I made a necklace for you out of these flowers. Come here, pet, let me put it on you."

"Please, let's go back now, Charlotte."

"Oh, fiddle, can't a body have her own way sometimes? Pa would have me busy every minute of the day in the house." She flung a pebble into the stream with force. "Well, I won't, Annie! I just won't stay home all the time. I hate woman's work with a passion. It's horrid, tedious, dull. Now, if Pa would let me help him in the store, I'd be ever so smart, quick as a whip to please him."

"You know he doesn't want you in Nauvoo with all the wild talk in the air about killing the prophet and the other leaders of the church."

"Oh, talk! That's all it ever is, talk! Nobody is going to kill Joseph Smith. It's been tried too many times, and he always escapes. Besides, that's what the Legion is for, isn't it, to protect the prophet? Pa just doesn't want his womenfolk seen in public. I think he'd like to throw a blanket over our heads every time we go out. The only man he allows to see me at all is Orin. I guess Pa figures I'm safe with him."

Annie laughed at her sour expression. "Well, Papa may rest his mind about it, but it's Orin I'm worried about. Aren't you ever gonna marry him, Charlotte? It would be the kindest thing you could do, you know. Then he might get used to looking at you and not faint dead away every time you talk to him." She fell backward on the grassy bank in a mocking faint.

"It'd be the unkindest thing I could do! He might stay that way until someone thought he was dead and buried him alive. No, I'm not gonna marry him and have him sent off on some mission two months later, like most of our men are. Besides, he won't marry me while I'm still a Gentile." She leaned over close to Annie's ear, "And truth to tell, that's as good a reason as any for not getting baptized."

"Charlotte, please, let's go back," Annie pleaded. " Maybe Esther can still get the grinding done before dark, and Ma won't be so mad. You know how she is. She can find a million things to keep you inside for days."

"I guess you're right, Annie. I just hate to be indoors in the spring." Charlotte heaved a sigh as she got to her feet, shaking off droplets of water. "I'll never be a lady and do needlepoint placers and crochet lace tablecloths. I don't see why I have to learn. It takes too long. There are so many better things to do." She stopped in the middle of the meadow, plucking up a buttercup to brush her cheek.

"Look, Annie, see how they kiss me. They smile at me and call me to come out and play. Maybe I'll just make me a little house out here in the meadow and live here."

She grabbed Annie's hand, and they ran through the tall grass, then jumped up on Esther's broad back. Charlotte sat in front, and Annie clutched her waist from behind. As they left the meadow behind, picking their way through the trees, Annie whispered in her ear, "Don't marry Orin, Charlotte. I'd be so lonesome without you."

"I'm not marrying anyone, pet. I haven't seen a single man that strikes my fancy. I'll know him when I do, though. He'll be dark-haired and handsome and will love me passionately."

Orin Southam was a Seventy in the Church of Jesus Christ of Latter-Day Saints. It was a rather curious calling for him, as it required him to be bold in teaching the gospel to non-members. Orin was a quiet man, a school teacher by profession. More scholarly than worldly, he abhorred going into business with his merchant father, so, instead, he was sent to Philadelphia to school and soon determined that he would like to spend the rest of his life studying the great philosophy of the ages. That was the reason he stopped on a street corner, one windy November day, to listen to two young men dressed in dark suits and long black capes exhort a meager crowd of people to read a new religious publication called the Book of Mormon. They preached to him out of the very scriptures he was steeped in. They taught doctrine unlike any other he had ever heard, advocating a preexistent life before man came to earth, preaching a heaven where all men received glory according to their works and faith on earth. They taught that he was a literal spirit child of God and challenged him to become perfect, thus becoming like God. Finally, they bore their simple testimonies of a modern day prophet, a young man named Joseph Smith Jr., and modern revelation from God. Three days later, Orin did the only rash thing he had ever done. He left his academy, his parents, and journeyed to the distant outpost of civilization—Nauvoo, Illinois—to meet the young prophet that taught such doctrine.

In Smith, Orin found the philosophy of life that appealed to all his reason and intellect. He also found his idol. He came to love him as he might have loved a brother. Some years later when Joseph was murdered at Carthage jail, Orin cried aloud in anguish, "I should liked to have died by his side."

But Charlotte O'Neill never saw the qualities of gentleness and loyalty in Orin that won him affection and respect in his circles. She saw only a young man, slight of build, dressed in an old, threadbare, school teacher's suit, embarrassed to meet her eyes, and too polite to reprimand her when she taunted him.

He had come today to ask her to go with him to the Church social on Saturday afternoon. He was waiting in the front room, chatting easily with Margaret when Charlotte and Annie rode up. After re-hitching Esther to the flour mill, Charlotte went inside.

Orin stood and bowed slightly, growing pink as Charlotte bounced up to give him her hand. "G . . . g . . . good afternoon. You're looking w . . . well today," he stammered.

Charlotte glanced at Annie, sharing a slight smile with her. She glanced at Margaret and swallowed the smile. "Why, thank you Orin," she said as demurely as she could. "You say the nicest things. What brings you out here on such a fine day?" She began to enjoy playing a part even if it was only Orin who was impressed.

"I, uh . . . well, it's the social . . . the Church social . . . I, uh." He looked at Margaret for support. "Yes, uh hmmm." He cleared his throat. "It is the Church social I meant to speak with you about." There was a pause while Charlotte looked at him expectantly, waiting for the inevitable. He pushed his thin hair back nervously and glanced around at Annie watching them, unabashedly interested in the conversation. "Charlotte, do you think we might walk about for a bit?"

"My goodness, Mr. Southam. I declare that's an unwelcome suggestion to a young lady who has just finished a most tiring ride." She sat heavily down on the window seat.

"I'm sorry," he stammered. "I only meant . . . well, I wanted to speak with you privately . . . begging your pardon, ladies." He bowed to them dramatically.

"It must be a matter of great importance to require such privacy," she said archly. "Have you spoken to father about it . . . ?"

"I think it would be quite all right for you two to take a short walk," Margaret interrupted her daughter, impatient with Charlotte's teasing. "And I don't think that Charlotte is that tired."

Orin breathed a quiet sigh of relief, thanking her with his eyes. Charlotte extended her hand in a most ladylike fashion. Orin helped her from her seat, and they walked outside to continue their conversation where, at least, the presence of others wouldn't magnify his embarrassment.

"Well, have you?" she asked.

"Have I what?" he replied.

"Spoken to father about your important proposal."

"Well, it isn't really that serious. I only meant to invite you to the Church social this Saturday."

"Oh fiddle, that! I thought you were about to make a proposal of marriage. Just to ask me to the Church social, you could have ridden up and shouted it from your horse and been on your way in a few minutes. No, I am not intending to go the the Church social."

"Please, Charlotte, It'll be such a good outing. Almost everyone in Nauvoo will be there, even the prophet."

"Why should I care if Mr. Smith is going to be there? You sound as though I have a flirtation going on with him." She loved to tease him. He suffered so exquisitely from it.

Orin was blushing profusely, and his chin was inches away from his chest. Almost inaudibly he said, "You know I didn't mean that."

"If you didn't mean that, what did you mean? Certainly Mr. Smith nor anyone else in the Church could care the slightest whether I am at the social or not. After all, I am just a Gentile, even though my family may be 'saints.'"

"That's easily remedied. It would be the greatest honor to baptize you."

She touched his cheek gently with her fingers, pitying him for a moment. "Yes, Orin, I know." He raised his head at her touch and looked full into her eyes for the first time that afternoon. There was such sweet love in his eyes that Charlotte felt herself threatened. So she withdrew again behind her banter.

"But once baptized, soon married. Once married, soon pushed aside for other wives. I'd rather not have a husband than to have to share him with another woman."

He stopped and took her hand tightly in both of his own. "Charlotte, if you would only consent to marry me, I promise you I would never, never take another wife."

She was coquettish, playing with the lapel of his coat. "Not even if your prophet ordered you to?"

He paused for a moment, reflecting. "But he wouldn't, not if you were so much against it."

She stamped her foot with irritation. "Oh, he would, too. Everybody knows he believes and lives this 'celestial' law himself. And if he commanded you through his 'revelations' to take other wives, your promises to me would be easily forgotten."

For that he had no answer, because, while he loved her more passionately than he could admit, he also knew that Joseph was a prophet. Why Orin loved Charlotte he could not have explained. It was, perhaps, the fatal attraction of the gray moth to the flame.

"But will you go to the social?" he was persistent.

She thought for a moment and decided it was better than needlepoint. At least she would be outside. "Oh, all right. If Papa approves, yes, I'll go. It's better than staying home. You're really very sweet, Orin." She enjoyed watching him blush.

"Fine," he whispered in relief. "I'll be here at noon to pick you up. Brother Taylor is lending me his carriage as he is going with Joseph and Emma. So we will have a fine carriage to ride in."

"Oh, well, why didn't you say so sooner. I'd have told you 'yes' right off," she laughed. And he laughed with her, too happy to care about her teasing.

It was Nauvoo, Illinois, 1844, as pretty a city as you could find west of Philadelphia. Bounded on three sides by the Mississippi River, it had been a lowland, swampy and unhealthy only three years before. Joseph Smith had selected the site as the place to build a refuge from the persecutions that had beset the Church from it's birth. They had drained the marshy land and proceeded to erect a busy, commercial city that rivaled Chicago in population and, certainly, in aesthetics. The streets were broad and laid out to the square. The Mormons cultivated their lots with all manner of vegetables and fruit trees, as well as banks of bright flowers. It was a tidy community, kept clean by industrious, conscientious people. They were poor only because of circumstances, many of them coming from the hard-working middle class of the eastern states and of England. They were self-sufficient in black-smithery, milling, education, printing, and, of course, farming. Joseph set the example in education by being actively engaged himself in learning German and Hebrew and in teaching religious principles in his School of the Prophets.

The pride of Nauvoo was the great stone temple inching upward to the sky. Built of pale gray limestone, it was one hundred and twenty-eight feet by eighty-eight feet around the base. It had a graceful elegance and pristine beauty that belied the rough frontier towns now bringing civilization to the bulging borders of the United States. The main building was three stories in height, and the towering spire would eventually rise three stories beyond that. Windows lined each story in rounded Romanesque style, lending symmetry and ornamentation to the building. It was the pride of all Nauvoo because all Nauvoo was personally involved in the building of it. The men devoted a full tithe of time to working on the structure; one day in ten was the least offering. Many devoted more hours than that. The ladies contributed food to the workers, linen, crocheted doilies and all manner of fine sewing for the veils and clothing to be used therein.

In such a damp spot, Nauvoo was bound to be green and verdant, particularly in the spring of the year. On the day of the Church social Charlotte was in high spirits. The early May sun painted the city in a bright green frock, and Charlotte decided she had been right to accept Orin's invitation. She was ready and anxious to be gone when Orin trotted up with an unusually fine horse and buggy. They packed the large picnic basket into the back of the buggy, securely lodged it beneath the seat, and smartly trotted off.

As the people gathered beneath the trees chatting and visiting, the Nauvoo Legion put on a demonstration of their marching drills. The military band played often during the day, enlivening the party with marches, waltzes, and patriotic music. Today Orin was at his best. He was proud with Charlotte at his side, and seemed determined to show her off. It was almost as though he was saying to her, "See, I'm not so backward, and the Mormons are not stuffy either."

The crowd formed their squares all over the grassy meadow, dancing to the vigorous music of the band. Charlotte and Orin had danced two sets, laughing and moving with the music. They were standing on the sidelines of the crowd, catching a breath and fanning themselves, when Charlotte heard a deep voice. She turned to seek the owner and recognized at once the General of the Nauvoo Legion, Orin's adored prophet, Joseph Smith. She had never been so close to him before. She saw the crowd part for him as he moved amongst the people and perceived the magnetism that drew people to him.

He shook many hands before he reached them. Charlotte glanced up at Orin and saw him anxiously anticipating a meeting. Then Joseph was beside them. He clapped his hand on Orin's shoulder. Orin immediately turned a perceptible shade of pink and seemed to grow two inches before her very eyes.

"Brother Southam, I don't believe I've ever seen you have such a good time. Brother Snow says you are one of our hardest workers, and I believe he's right. I've been noticing the long hours you put in on the temple. It's good to see you having fun." Joseph turned his attention to Charlotte and smiled. "Is this your young lady?"

"Yes sir," Orin replied proudly. "Well . . . that is . . . actually I'm hoping . . . uh, this is Charlotte O'Neill, President Smith. She's Patrick O'Neill's daughter."

Patrick had spoken to Joseph just the day before about his daughter, telling him a little of his concern for her hot-headed, stubborn ways, and asking him if he could find time to speak with her.

"If ye could only talk to the lass, President Smith, I know she could na' resist the spirit. She's a good girl, and the apple of my eye, I must confess, but I can na' talk sense to her."

Joseph had promised him he would do what he could, and for that specific reason had now approached Orin and the pretty, red-haired girl he recognized as Patrick O'Neill's daughter.

Charlotte did not want to be impressed. Nevertheless, Joseph was a rather commanding figure, particularly dressed as he was now in his white General's uniform with the gold epaulets. Charlotte automatically responded to the introduction with a slight curtsy, then quickly caught herself and said, "How do you do, *Brother* Smith."

Orin colored a little more at her audacity. To him, Joseph was 'Your Highness.'

"You've been enjoying the dance I see." Joseph was not perturbed by Charlotte's presumptuousness, but chatted with them cordially. "Perhaps Brother Southam would permit me the pleasure of a dance with you."

Orin quickly agreed, and before she knew it, Joseph had led her into one of the sets of dancers. He was a good dancer, moving smoothly to the music and enjoying himself famously. He laughed aloud as he twirled her round and round. His eyes were sparkling as he guided Charlotte in and out of the dance patterns. The music grew faster and faster at the end. Couples were whirling, and Joseph was twirling her about so fast her head was fairly swimming.

Green eyes sparkling, her face alive with enjoyment, she pleaded, "Stop, oh stop. My head is spinning!"

Joseph slowed and, in a moment, it was over, the music stopped. People were slapping their knees and catching their breath all around.

"You're a wonderful dancer, President Smith," she admitted, her hand on her chest, her breath coming in short gulps.

His chuckle was deep and full of merriment. "I'll say the same about you, young lady. Your little feet kept right up with my big strides."

He was escorting her back to Orin's side, when suddenly she wondered why she had always disliked him so. She had never thought him human. But he was. Then her defenses came up again. Too human, she thought. "Do you suppose the prophets of old danced and wrestled and told jokes as you do?" she asked him boldly.

His eyes never lost their smile, although his eyebrows raised slightly. "David danced. He also sang and played a lyre. Elijah taunted and played jokes on the priests of Baal. Jacob wrestled with an angel. Are you wondering if I am in good company there, or if I am just a young reprobate?"

Orin blushed at Charlotte's frankness, and she colored a bit herself. Lowering her eyes she replied, "No . . . I . . . I just wondered. You seem 'human' to be a prophet of God."

"Perhaps you mean 'so common'." Now he was watching her intently, studying her. "Others have accused me of that. I do not try to be what people think of as a prophet. I try to be what God wants me to be. Perhaps I don't seem particularly 'holy' to you. Perhaps I'm not! I simply know what I know and do what I do. Sometimes God gives me wonderful answers. Sometimes I make mistakes and wonder why He let me. Does it offend you for people to call me a prophet?"

Charlotte was uncomfortable now. She had found herself enjoying this man that she wanted to dislike. She had expected to find him conceited and arrogant. He was not. He was warm and fun-loving, and he had an instant rapport with people, even her.

She looked boldly into his blue eyes. "It has at times."

"I see. And now?" She could hardly look into those eyes for long. His penetrating gaze made her very uncomfortable.

"Not so much," she admitted quietly.

"Good. I don't mean to offend anyone. I frequently find, however, that people are offended only if they choose to be. I don't ask anyone to accept me as their leader. They simply choose to do so or they don't." He paused, then asked as she looked back up at his face. "How do you choose, Miss?"

"I . . . I don't know. I guess I haven't chosen yet." She was stammering. He was smiling.

Orin was very much enjoying Charlotte's discomfort, for she was usually the one with the upper hand. But she had met her match in Joseph Smith, and all her flippancy seemed out of place. His direct manner, coupled as it was with courtesy, seemed to bring out a certain humility in her. Orin knew Patrick would be pleased when he heard about all that had happened.

Bowing slightly from the waist, Joseph said good-bye to Orin. Just as he was backing away, he pointedly said to Charlotte, "Choose soon. You can't afford to wait much longer." Then he was gone, shaking hands, slapping men on the back and moving amongst an adoring crowd, back to Emma's side.

It was a wonderful social. Spirits were high, the day was warm, and spring promised grand things. There were foot races, in which many of the apostles joined, and a wrestling match, during which Joseph threw every man that came up against him, including Hosea Stout, a bull of a man. Orin joined none of the games, but led Charlotte in many of the dances, waltzing her around and around beneath the shade of the trees. He was a good dancer and graceful, but, even with the events going on, Charlotte never felt really comfortable. Many of the ladies asked after her mother, yet the conversations dwindled after a few exchanges. She was an outsider.

Lunch had been over for a couple of hours, the games were going on all about them, the band was playing, and Orin asked her for still another dance.

"Oh, Orin, couldn't we go for a ride?"

"Why, are you bored?" he asked anxiously.

"No. It isn't that. It's only that I've danced enough and pretended enough for one day."

He was disappointed. He had thought she was having a good time. Unobtrusively, they gathered together their basket, the tablecloth, the remainder of their lunch, and quietly took their leave, saying good-bye only to John Taylor. It was too fine a day, Charlotte felt, to spend her precious time being polite to people she had nothing in common with. She was much more at ease with Orin. Now, in the late afternoon sun, swaying slightly in the buggy behind Taylor's horse, she was relaxed and pleasantly warm. Once she had him stop, so she could pick a bunch of flowers growing beside the road.

Charlotte hummed lightly in a high, thin voice, as they trotted along. "Orin, someday I'll have a fine horse, as fine as this one. You know, I admire a beautiful horse, almost as much as you admire the prophet. I'll bet I could train any horse in the world to do just what I wanted. It only takes a little love and patience." She sat back lazily. It was a good day. She was as happy as she had ever been. Moreover, she felt herself mellowing toward Orin. She stole a glance at him—not bad looking, though he could use a little sun on his face. His eyes were large and fringed as delicately as any girl's, and his brown hair, with the encouragement of the sun, could be an attractive blonde.

"Orin, I have an idea. Let's drive over to the meadow. It's my favorite place, and it has a stream we can wiggle our toes in. I'm awfully warm."

So she gave him directions, and, in a few minutes, they came out into her meadow. They draped the reins of the horse over a bush, and left the animal to scrounge amongst the tender flowers and grass. She grabbed his hands and started to bob and dance around him, turning him in circles until he begged her to stop. It was good! It was the first fun they had had together. She tossed flowers over his head. On the bank, with bared feet dabbling in the stream, she teased him continually with grass in his hair and by tickling his face. She pulled out the pins that held her hair in the conventional bun, and down it fell. She shook it out, glinting red and gold, like a summer sunrise. It almost reached the ground as she sat.

Desire welled up in Orin, and he reached for her, hoping to be permitted a kiss. But no, she was enjoying her power and meant to keep it. Charlotte jumped up and promised him a kiss, if he could beat her to the blackberries growing thick downstream about a hundred yards. He won the race, but she made him race back to the stream. By the time they reached the water, they were both hot and perspiring.

"A kiss, for such a slowpoke? Ahh, no! I won the race back. Well, so I did trip you. No matter. Still, I won."

A delicious thought occurred to her. "Orin, aren't you a bit warm? I certainly am. I have an idea." She turned quickly about and confronted him. "Let's go swimming."

"But, Charlotte, I . . . I haven't my swimming outfit with me. Neither have you. Besides, gentlemen and ladies don't go in bathing together."

She laughed at his consternation and said, "Have you never been swimming without an outfit? It's much the best way."

His face was a pathetic study of shock, disbelief, and confusion. Tempted? He wasn't the least bit tempted! He wanted Charlotte as a wife. Besides, Orin would have been as embarrassed to see, as to be seen.

Charlotte knew there wasn't the slightest chance he would take her up on it. She was safe, and so she teased him the more, chiding him for his timidity, trying to tempt him with the coolness of the water. The more she entreated him, the more desirable the idea became to her. At last, she told him, "All right, if you are afraid, I'll go swimming alone. You can sit and watch." He was just as horrified at that suggestion. Indeed, she giggled to herself at the glistening drops of sweat her suggestion caused to spring up on his forehead and upper lip.

"Oh, fiddle, Orin, you're just no fun at all. All right, go over there behind that bush, and read your poetry whilst I go swimming."

"You're not, really, are you Charlotte?" he asked imploringly.

"Certainly, I am. Except, wait. Let me get the tablecloth we used, so I can dry off when I'm done." She dashed over to the buggy and pulled out the tablecloth, waving it in the air as she brushed past him on her way to the stream. Helplessly, Orin sat down behind the bushes that grew a hundred yards away from the stream. Determinedly, he turned his back, as she disappeared behind a high stand of grass, by the water's edge. He wished that he knew how to appeal to the more refined instincts of the girl. He felt himself lost and ineffective in winning her.

From the bank of the stream, with the line of bushes that bordered it, she could scarcely see him at all. Every now and again, he moved minutely, and she could glimpse the roundness of one shoulder. She paused for a few minutes, wondering if she really should. It was very naughty, she knew. Still, what harm could it do? If it were anyone but

Orin, she would never consider it, but she knew she was as safe as if it were Patrick O'Neill sitting there. Nervously, Charlotte glanced around, a bit more reluctant than she had let on. Finally she began undressing, but as she reached her chemise and pantaloons, she paused, then decided she was already daring enough.

Leaving the rest of her clothing neatly folded up on the bank, and glancing quickly at the bushes that concealed a very skittish Orin, she waded out into the water. It was chill, spring water. Had it been deeper, the May sun would never have warmed it up enough, but as it was, Charlotte quickly became used to it. Soon, she was gliding smoothly about the small perimeter of the little pool she had dammed up every year since they had moved to Nauvoo. The late afternoon sun shone hotly. Charlotte lay back in the water, squinting up at the blue expanse above her. She had twisted her hair back on top of her head helter-skelter and pinned it with her hair-comb. Still, shorter strands of it fanned out in the water around her, clinging wetly to her neck when she paddled about. She dipped her face in the water, shaking the droplets of liquid away. Soon she was singing to herself some old Irish songs she had learned from Patrick.

After a while, she called out, "It's grand in here. You should have come in."

At that moment, she heard a deep voice, "Perhaps I will."

Charlotte's eyes shot open, as round as saucers. Her hand flew up to her mouth to stifle a scream. Thoughts flashed through her mind so fast she hardly had time to realize each one completely. One thing was clear, she couldn't scream and alert Orin. Her reputation would be lost forever.

The man stepped out from behind the pampas grass and bushes, where he had been seated for better than a half hour, chewing on blades of grass, watching the scene going on across the steam. He was a big man. She judged him to be about half a head taller then her own father. His hair was coal black and almost as shiny. Dark eyebrows and beard were thick but neatly trimmed. He wore work clothes, but his boots were of fine leather and the best workmanship. His hat was felt. He held it clutched in his hand, as he started to move closer to the north bank of the stream.

She panicked and started waving him away, but he didn't stop. She glanced over her shoulder toward Orin. He was too far away, thank goodness, to hear and was still sitting, dutifully gazing in the other direction. The stranger advanced slowly and silently, until he had squatted down right on the bank, grinning at her. She had moved quickly to the other bank and scrunched down in the water, until it touched her chin.

"Maybe I just will," he said softly. "It looks very inviting in there."

"Don't you dare," she said, vehemently, yet taking care not to speak too loudly. Now that the surprise had worn off, she was becoming angry. "Ye ought to be ashamed of yereself. Just like a naughty little boy peeping on the girls."

"A naughty boy and an equally naughty girl." He smiled at her, in complete unconcern for her discomfort.

"Go away," she whispered. "Ye had better go, before me fiance discovers ye're here and gives ye the thrashing of yere life."

He chuckled quietly at that, and, again, she glanced in Orin's direction to make sure he hadn't heard. The stranger sat with his hat pushed back from his forehead, chewing thoughtfully on a piece of grass.

"I tell ye, ye had better get away as quick as ever ye can. He has a mighty hot temper."

"Why not call him, then?" He guessed that she wouldn't. If she had been going to, she would have done so by now. He guessed, too, that she was embarrassed to have the fellow know she had been observed.

When she remained silent, he taunted her, "Aren't you going to call him? No, you're not, or you would have already."

She blushed with anger. He continued, "You know, I've never seen one of those mermaids that sailors talk about, but you must come pretty close. Swim some more for me, little mermaid with the red hair."

Embarrassment took a second place to her hot temper, and she threw some water as hard as she could. A few drops reached him. He laughed at her again. Then she reached down to the rocks that were embedded in the soil of the steam and grabbed one. Still immersed in the water, she raised her arm and chucked the stone. She was a fair shot. It hit him on the shoulder.

"Listen, my naughty little girl," he warned her. "I could have you out of that water and turned over my knee in a wink. Naughty little girls deserve paddling."

She knew he could, indeed, and controlled her temper with some effort. She decided to wait him out. She did some figuring too, and decided that if he had wanted to do that, he would have done so already. He was just having his fun. Sure it was at her expense, but she had more or less invited it. After a few minutes she whispered, "Won't you please go away? I've been in here long enough, and I'm getting chilly. Besides, my folks will be wondering where we are."

He turned that over in his mind, and finally stood up. "Okay. I've seen enough. Thanks for the show, and by the way, little mermaid, you sure are the prettiest thing I ever have seen in Illinois." Bowing gallantly, he whispered with a mocking grin, "If you would take your bath here every day, I would leave my ranch and my horses and just set myself down beside this little stream. Good-bye, Red, I hope to 'see' you again some day soon." She threw some more water at him in sheer frustration, as he ambled off.

Charlotte watched him disappear into a clump of trees a little way upstream and then heard the soft clip-clop of hooves. After a few minutes, she decided she was safe.

"Charlotte, you should be getting out, don't you think? We must be getting back, or your father will take me to task for keeping you out so long." Orin was calling to her, with his head determinedly turned away.

"I'm getting out, Orin. You just stay there." She wrapped the tablecloth quickly around her as she emerged from the chilly water. Shivering now in her wet underclothing, she was sorry she had been so head-strong. She hurried getting into her slip and petticoats and dress, fastening all the buttons with chilled fingers. Finally she walked over to meet Orin, a much more subdued girl than when her swim had begun.

Jack Boughtman had his last glimpse of her as she rode away with Orin in John Taylor's buggy.

CHAPTER 2

His only love in life was his horses. He had no family, his mother and father having both died in a flash flood when he was thirteen years old, and his only brother was a frontiersman, whom he hadn't seen in years. He had no sweetheart, only women who were paid to share their favors. He had no close friends. The men he associated with in the Masonic Brotherhood were business connections, whose company was good for his business.

Jack Boughtman was more fortunate than many men, in that his business was also his play—horses! He broke the wild ones, birthed the foals, and trained them to suit the customer. Jack owned perhaps the largest horse ranch in Iowa, and it was situated just outside Montrose, across the river from Nauvoo. Business had always been good. His reputation for good horseflesh brought business from as far away as Quincy, and now with the Mormons coming in by the droves, he could scarcely keep ahead of his orders. Only just recently had the Mormon orders begun to drop off. The only explanation that came to mind, when he tried to understand the reversal, was that he had crossed their dear 'prophet' Joseph.

Smith had come to him a month ago wanting to purchase a pure white mare. Boughtman had a beauty. She had been bought from a dealer in Missouri three years ago, and he had bred her twice. Smith wouldn't wait for another foal; he wanted the mare. So Jack set a price on her—a steep one—but, what the devil, Smith could afford it. After some dickering, the price was agreed upon, and Smith was to come after the horse in a week. A few days after the deal had been made, a man rode in from Chicago with the express intent to buy that mare, no matter what the cost. Jack saw the chance for a real killing, so he set a price of three hundred dollars, and the Chicago man bit. Boughtman's biggest mistake was underestimation of the Mormon prophet. He considered Smith a pantywaist and a religious fanatic. He had never seen him angry. When Joseph came after his horse at the end of the week and found her gone, sold to another, he turned on Boughtman with an indignant anger that brought the workings of the ranch to a standstill. Joseph Smith was a big man, as big as Jack, and seemed to grow even taller with righteous indignation. His blue eyes pinned Jack and seemed to pierce through to Boughtman's heart of hearts. The prophet's voice rang out for all to hear.

"Jack Boughtman, you have tried to cheat me. You will try to cheat God, but as sure as I stand before you, Satan will have your soul if you do not change your ways, and him you cannot cheat."

No man had ever spoken to Jack that way. His height and brawn had always been enough alone to influence respect, and his cold, brutal temper was widely known.

The hired hands saw his fury begin when Smith railed at him. They privately thought the prophet was in for a skinning, but he jerked off his coat, and the two men wrestled for a good half hour. Jack was forced to admit to himself, after a while, that Smith was the best fighter he had ever come against. When, after thirty minutes he was tiring and the pantywaist prophet hardly seemed winded, he grudgingly promised Smith his pick of any two horses on the ranch. Joseph picked out the two best mares on the farm, but, as he rode off, he turned back to Boughtman.

"Mark my words, Boughtman. Satan is a hard taskmaster. You'd do best not to traffic with him. Hate me all you want, but you can't cheat me, and you can't beat me. Even if you should kill me, even then I would triumph."

Jack watched him ride off in his buggy, trailing the pick of Boughtman horses, and swore a curse underneath his breath. "Damn you! We'll see about that."

So, now that business was falling off, Boughtman attributed it, and rightly so, to the influence that Joseph had with his people. Word got around quickly that Jack was not to be trusted, and the Mormons began going clear down to Carthage to buy horses. Still, even without the Mormon trade, Jack did a brisk business, and was careful never to cheat his Masonic brothers in his dealings. It was not from any moral sense but a practical business mind. He knew where his bread was buttered.

Jack Boughtman was not alone in his hatred of the Mormon prophet. The other leading citizens of Montrose, Carthage, and Flint began to grow apprehensive about the influence of the man and the prosperity of the sect. Nauvoo was the fastest growing city in Illinois. They used their political block of votes to affect the policies they believed in. When the Presidential elections began, Joseph Smith and Sidney Rigdon mounted their own platform and began sending their force of missionaries to spread the political 'truth', as well as the spiritual. Resentment grew and persecution increased. Adversaries ferreted out the discontented among the flock, using them to spy on Joseph and spread dissension in Nauvoo. They were trying to get a newspaper going that would cast aspersions on Joseph's character and thwart his plans. As yet, it had not printed a single issue. They were timing it very carefully.

There had been isolated cases of harassment of the farmers that had land on the outskirts of town. No one had been killed, but there had been plenty of threats made, barns burned, and the temper of the people began to be on edge. Boughtman had been in on the harassment. He enjoyed power and the respect that fear brought. He laughed at Charlotte's threats that her fiance would thrash him. He knew Orin— at least by reputation—and knew that he was as likely to fight as a month-old kitten. Still, it was the fire in her green eyes that appealed to him. That she was high-spirited made her the more appealing. He loved a challenge and rarely had one, because most people, women especially, were a little afraid of him. They instinctively sensed his capacity for anger and backed away from him, despite his growing wealth and influence.

So, he had watched her dress from a few yards further away, wondering who she was and what such a jewel was doing amongst the drab Mormon women, forever dressed in their modest homespun dresses. It didn't take him long to find out. He frequently went to Nauvoo on business, and, secretly, to stir up opposition to Smith and the other Church leaders. Less than a week later, he stopped in at O'Neill's store to pick up some supplies he had ordered.

Charlotte had begged Patrick and wheedled him, until he finally had given in and promised her she could work in the store, if she would mind her manners and her tongue. She was now busily reorganizing the store to show off the items most appealing to the women who came in. She had just finished rearranging the bolts of cloth and was admiring her work, when she heard a deep voice behind her.

"Well, lookee here! If it's not my red-headed mermaid. You're even pretty with all your homespun 'finery' on."

Charlotte spun around, recognizing the voice immediately. "What are you doing here? And who are you? You'd better swallow that kind of talk. My father has an Irishman's temper."

He leaned up against the counter, surveying her. She shivered once, then turned up her nose at him, and swung through the doorway into the offices in back, where her father was doing some long overdue book work.

"Papa, there's a 'person' to see you."

Patrick looked up, hearing her disdainful tone, and said, "Keep a civil tongue in yere head. Our customers are gentlemen and ladies to you. Don't ye be offending them. Who is it?"

"I'm sure I don't know. But I do know that he's hardly worth the title of gentleman."

"Why?" Patrick was becoming suspicious now. "Has he been out-of-the-way with ye?" He leaned way over and squinted, in order to see through the crack in the curtains that shielded the door. He caught a glimpse of black hair and mustache.

"Oh yes, Mr. Boughtman. Um, well, it's no wonder you do na' think much of him. Though he's never been amiss with me, there's rumors a-plenty about him." He left his books and put a smile on for his customer.

"Yes, yere supplies have come in. I think the leather is not so good as we've been getting. Still, the tobacco smells fine enough for a politician." Pat brought out the supplies he had laid aside for Jack, and all the while, Charlotte stood, half hidden, behind the curtains.

"Who's your helper, O'Neill?" Boughtman asked.

Pat was reluctant to make the introduction. "Why, uh, this is me daughter." He hurried on. "She is just helping for a time."

Jack was staring appreciatively at Charlotte, "Does she have a name?"

Pat glanced at Charlotte, "Come here, girl." He put his arm around her waist, half-affectionately, half-protectively. "Mr. Boughtman, this is me daughter, Charlotte. Charlotte O'Neill." He laughed sociably. "She favors me. Don't ye think?"

"Hell no! She's prettier than you, by a damned sight, O'Neill! Where'd you get such a gorgeous critter? She's finer looking than any of my mares, and I've got the best in Iowa Territory."

Charlotte was blushing now, and her eyes were turning emerald. As her temper flared, she fell into Patrick's brogue. "Ye might have the best horses, Mr. Boughtman, but yere manners would na' bring ye a shilling. I do na' mind bein' compared with yere mares, but I won't be called a critter."

"And spirited, too." Jack laughed.

"Aye, she is that, when she gets riled up. Mind yere own manners, lassie," he cautioned her.

"Well, O'Neill, figure up how much I owe you, and I'll pay you today."

Hesitantly, Pat made a move toward the doorway, looking back and forth from the girl to the man. "All right, I'll be just a minute. Charlotte get finished with yere work out here. My offices are in need of tidying. And mind yere manners, girl."

"Charlotte O'Neill." Jack said in a low voice. "Kind of a prim and proper name for a spitfire like you."

She didn't answer him, but bent over the threads and sewing goods. "Let's see. What's a better name for you? Kind of saucy for a girl, ain't you? I'd have named you Charly! That suits you better I think."

She reacted swiftly to that. "Charly! What an awful name. It just matches your manner."

"Well, you sure ain't no prim and proper lady, to be calling you Charlotte," and his voice got prissy on the word. "No sir, Charly's the name for you. It just matches the red in your eye when you're all fired up."

Now she glared up at him. "Ye are the most impudent, ill-mannered excuse for a man I have ever met." She started past him to the curtain. "Good-bye, Mr. Boughtman. I sincerely hope we never have occasion to meet again."

She didn't get quite past. His grip on her arm stopped her, when she was directly in front of him. His hand was hard and his grip was tight, and for a moment, she was afraid of him. When she looked at his face, though, she was surprised. It had softened, and there was no derisive smile now. There was a longing, almost a gentleness, that confused her.

"We'll meet again all right. I told you once that I would leave my ranch and my horses for you, and I think you really could tempt me. But, perhaps, I might tempt you, Miss O'Neill. My ranch and horses are worth seeing. Forgive me my ill manners. I'm not very polished around the ladies. I save my sweet words for my mares. Perhaps you could teach me some appropriate ones.

Wary as a cat, she held herself tense. "I'm sure the taming of you would be more than I could manage, Mr. Boughtman."

"Jack. My name's Jack. The taming might take a bit of doing, but the trying could be fun." His hand was gone from her arm, but still he held her with his eyes. He spread his hands out, in an open gesture. "Just tell me what you like—anything you'd like for me to do—and I'll start right away, trying to please you."

"The first thing would be to use appropriate language around a lady. If you can manage that, you might then go to work on the way you look at me. You look at me, for all the world, like a hawk circling a field mouse."

He laughed and replied. "I'll try little lady, but I'm afraid I'll never manage those adoring timid glances of your schoolmaster fiance. If he is your fiance?"

She blushed at her lie. "Well, no. He isn't, actually, though he has asked for my hand often enough. Now, I really must go back and help Papa."

Again his hand stopped her, but gently this time. "Say, Charly, let's us be friends. I won't be rude anymore. I'll be a model gentleman from now on."

She started to protest, but he cut her off. "I know you don't want me to call you 'Charly'. All right, when there is anyone else around, I'll be very proper and call you Miss O'Neill. But just between us, I think 'Charly' suits you better."

From that day on, he called her Charly, and she soon grew to like the name, secretly feeling that it did suit her better. Every time she said it to herself, she saw him smiling at her, and a delicious thrill ran through her.

Jack began to hang around O'Neill's store. Every day he found occasion to come into Nauvoo on business, and every day he wanted some item from the general store. He made it a point to be very polite to Charlotte and sociable with her father. At first, Pat was leery and only as polite as civility demanded, but, after a while, he began to let down and gossip with him on the latest developments of the Church, just as he would his Mormon friends. Sometimes Jack made the opportunity to talk with Charlotte alone, and other times he deliberately came in when she was not there. He would place an order with Pat, and then stand around, trading gossip.

Pat told him that the Prophet suspected a renegade printing press was trying to get started.

"No kidding? Somebody must not cotton to him, huh?"

"Oh, I don't think it's anything to be worried about. Just some apostates, lost souls if ye ask me, who have had the word, and then lost the spirit. Joseph'll handle it. We've had much worse things than that to persecute us."

"Yep, I 'spect you have. Aw, Joseph'll take care of 'em, all right," Jack said. "He strikes me as a man that could beat the tail off a bushwhacker and come up with nary a wrinkle in that ruffled shirt of his."

He laughed at that, with Pat joining in. Jack's laughter stopped suddenly, and his face sobered up. "You know, I wrestled with your prophet once."

Pat looked at him sharply, the word having gotten around. "Yes, seems to me I heard something about a run in ye had with Joseph."

Jack dismissed it with a wave of his hand, "Pshaw, weren't anything important. We was having trouble settling a deal over some horses. So Smith, he ups and takes off his coat and says to me, 'Jack, tell you what. I'll wrestle you for it.' 'It's a deal,' says I. Well, we wrestled for nigh on to half-an-hour, and I can tell you Smith is no slouch. No sirree. He like to had me a couple of times. Finest wrestling man I've ever come against, and I've wrestled many a good man." He nodded his head positively. "You know nothing wins my respect like a man that's not afraid to get his hands dirty. Yep, Smith's all right in my book!"

Pat listened carefully, matching up Jack's account with the rumors he had heard. Boughtman's manner was so congenial and matter-of-fact, Pat decided that he must be telling pretty near the truth. He was glad to see Boughtman was not so prejudiced against the Church leaders as some people said. Patrick began to imagine he might even be able to influence the man for good.

One sun-filled morning, Jack was waiting with a brand new buggy and his finest horse, when Charlotte came swinging around the first bend in the road from home to town. She was on her way to the store, and was so glad to be out of the house, finished with her needlepoint, that she was whistling.

"Never saw a lady that could do that. What 'cha up to, Charly, my girl?"

"Headed into town to the store, as if you didn't know, and I'm hardly your girl." She cocked her head at him, squinting up. "What are you up to, Mr. Boughtman?"

"I just drove along this way to see if there were any 'fair maidens' who might enjoy a ride with the finest horse and buggy in Illinois."

Charlotte smiled at him now. Then he quickly said, "But I don't see any, so I'd better be getting back." He gave a cluck to the horse, and the carriage moved a foot.

"Jack Boughtman, ye are a scoundrel!"

"Yep, so I've been told before. Come on, Charly, I guess you'll do, seeing as how there ain't no other fair maidens around."

She turned away from him and started on down the road, swinging her lunch basket and whistling a thin little tune. "Well, it's such a mighty fine day. I don't think I'd care to ride in a buggy, behind a nag, with an impudent man. Walking just isn't that bothersome to me."

"Got yer goat, didn't I?" He followed along beside her. "You know I sat here, just thinking of what I could say that would make your eyes green. And I says to myself, 'now the first thing a woman likes is to be called pretty, and the second thing she likes is to be seen in fine circumstances. Charly is a girl, just like any other, except for the fact that she is prettier.' So, I just worked the thing in reverse, and, sure enough, I got your goat. I'm a little disappointed in you though. I didn't think you'd be taken in so easy. I thought you were smarter than that."

"Ha! I'm smarter than you ever hoped to be and, certainly, more refined. Whatever made you think that a fine horse and buggy could give you the polish you'd need to be seen with a girl of quality like myself? I'm a lady, a businesswoman. I'm educated, and, like you said, I'm a far sight prettier than anyone else around here at the moment. So trot along. I'll just wait for my prince to come along and sweep me up."

He listened to her, smiling, then stopped the buggy, jumped down, and scooped her up in his arms, carrying her to the buggy.

"So that's what you want, to be swept off your feet?" He smiled into her eyes, his face only inches from hers.

"I said a prince!"

"I keep my kingdom's jewels at home, locked up in my tower."

"Is that all you have in the whole tower?"

"For now it is. But it's waiting for a fair maiden. What do you say?"

"Surely, I'm not fair enough." Her gray-green eyes were laughing at him. He held her easily in his arms, beside the carriage seat. She felt a shiver run through her, and she was sure he could hear the fast thumping of her heart. She remembered what she had told Annie once, that her man would be dark and handsome and would love her passionately. Jack was certainly the most exciting man she had ever met, as well as the most handsome, with eyes as black as night, and shoulders as wide and hard-muscled as iron.

"You can set me down anytime you want. That's what the seat is for." She was acutely aware of the warmth of his body and the strength of the arms that held her.

"You didn't answer me about the tower," he said.

"Nothing to answer. There isn't any tower, and you're not a prince, and I'm certainly not a princess."

He grinned at her, then half-tossed her into the seat. "Well then, woman, how about a ride behind a nag with an impudent man like me?" They both laughed at their silly play.

Cantering into town, Jack drew the horse up sharply beside her father's store. He climbed out of the buggy, graciously lending her his arm, helping her out of the carriage. Loudly he said, "Miss O'Neill it's my privilege to be of service to you." As they approached the doorway, he whispered to her, "I'll see you by the bend tomorrow." And so began their daily meetings.

On the first evening in June, just as night was coming on, Jack rode up to the house. His horse was in a lather, so he and Pat walked him out while they talked.

"What's yere hurry, Jack?"

"Oh, no hurry. I like to put my horses through their paces. This mare's been lazy lately, so I've had her out training." Still, he kept glancing around occasionally, as though he expected to see someone coming.

When the horse was quieted down, they went inside. Pat O'Neill's house was not fancy, rough plank-floors and walls, with logs overhead forming the ceiling. But, at the open windows hung curtains of soft white linen. Braided cloth rugs were covering the floor boards, and samplers that the women had made were everywhere to be seen. Margaret had lit the candles about the room, and the soft smell of wax burning was sweet in the yellow, shadowy light.

Dinner time at the O'Neill place was the focal point of the day. Patrick always had news for the family of all the goings-on in town, and recited, with perfect imitation, the conversations he had had with the townspeople that came into his store for their goods. Such a mimic was he, that he had the girls in convulsive laughter at times, and even Margaret could not dampen him when he was in really good form. She would scowl at him and admonish him to be more respectful of their neighbors. He would only go on more outlandishly than before. Tonight had been one of their more quiet dinners. There had been talk in town about mobs of men doing mischief and threatening Mormon farmers.

Jack and Patrick sat talking about business for half an hour, and Pat still had not called Charlotte in. After drawing Patrick out on what he knew about the illegal press in town Jack asked, "Is Charlotte elsewhere engaged this evening?"

Margaret's head lifted from her concentration on the shirt she was mending. She studied Jack seriously, and there was the slightest suggestion of a frown between her brows.

"No," Patrick answered. "I reckon she is setting in her room, waiting for me to call her out." He smiled. "Is it a coincidence ye have brought her into town the last four or five times?"

"What do you think, O'Neill? No, it isn't any coincidence. Does that bother you?"

Patrick looked at him for a full minute trying to make up his mind about the man. He was bemused. "Well, it doesn't bother me none, just as long as you know I won't have the lass marry anybody but a Mormon. And I'm sure ye would na' be thinking of anything less than marriage."

The frown deepened on Margaret's brow, and her lips pursed up in unspoken words. She didn't like Jack Boughtman. From her first sight of him as he rode in this evening, she had not liked him. There was an undercurrent of lightning that seemed to run through him, a sense of wary wildness that pervaded even the set of his shoulders and the way he turned his head. Jack Boughtman and Charlotte O'Neill together would be like lightning and the storm.

Jack smiled, "Are you now? No, you're not that sure, are you Pat? Well, don't let it worry you. You don't know me well enough to know that I do things the right way, or not at all. If I want something, I come right out in the open about it. I'd never play the sneak and ruin your little girl. I like her too well for that. I don't mind telling you that I like her well enough to make a wife of her, if she will have me."

Margaret stood up. Jack had forgotten the quiet, small woman, silently sitting in the shadows of the room. He could see at a glance the disapproval on her face and detected the unmovable quality about her. "So that's where Charly gets her backbone," he thought.

Margaret began, "Mr. Boughtman, Charlotte is not . . . " At that moment, Charlotte appeared in the doorway at the tiny bedroom.

Jack stood up. She was gowned in a dress Annie had just put the final buttons on moments before. It was yellow cotton, and it had six buttons up the sleeves that fitted tightly from the wrist to halfway up the elbow. The bodice was shirred with a rounded neck. Her waist was tiny and slim, and the skirt of the new dress fell in graceful lines about her hips. Pat looked up, and his heart melted, as it always did, at the sight of his favorite daughter. Despite the many angry conflicts, he adored this fiery girl.

"Charlotte, Mr. Boughtman dropped over to set awhile and talk. Would ye like to set with us?"

She smiled fondly at her father, grateful that he was not going to be difficult. "Yes, thank you Papa," she said, deliberately not looking at Jack. Yet, she saw him from the corner of her eye. He was silent for the first time, admiration plain in his eyes. Margaret saw it, too, and looked at her daughter. Charlotte had softened lately, becoming more pliable and agreeable. Margaret shook her head slightly. Could Charlotte be in love with this evasive, inscrutable man?

"You seemed to have removed your beard, Mr. Boughtman. I hardly know you without it."

"Oh, the beard. I . . . uh, someone said that Miss O'Neill didn't favor men who sported beards."

She smiled at him, pleased at his obvious play for her affections. "So you removed yours. How flattering! Is it business that takes you out this evening?"

That startled him, and he said quickly, "No! No business. I wanted to work my mare a little. She's been getting lazy. So, I put her through her paces, then let her out to see how long it takes to lather her. Not long I must say."

Charlotte had been making friends with his horses as he drove her to town each day. "Which mare is it?"

"Rose O'Sharon. She's a good breeder, and has some roan in her, but I've been neglecting her lately, and she's getting fat and lazy. Would you like to take a look at her?"

So the three of them went out to where the horse was tied. Margaret resumed her favorite rocker when they left, waiting for Patrick's return so she could voice her objections of the dark man, Jack Boughtman.

Charlotte made immediate friends with any animal. Horses were her favorites. She loved their graceful lines, thin legs, big soft eyes, and long noses that she stroked fondly. She even liked their pungent odors. She could get a horse to do almost anything for her. Even Esther would jump a fence for her, if she coaxed the animal softly for a few minutes. They stood talking horses for many minutes, stroking Rosie's nose, scratching behind her ears. Finally, Charlotte turned back to Jack, her eyes aglow, "She's a dear, so gentle and friendly. Look at her, she trusts me already."

"It's no wonder. Anyone as pretty as you could steal any heart."

Charlotte glanced at her father. He surprised her by nodding his agreement. "Tis true, ye are very pretty tonight, lass. Only take care to remember that 'pretty is as pretty does.' I will be seeing if Margaret be a needing me." He put his hand on Jack's shoulder. "See that ye keep in mind what ye told me this evening, for I'm trusting her to ye—for a little while, that is. Glad of yere company, Jack. Come back again."

They watched him go into the house, and all was quiet between them for a few minutes. Charlotte stroked Rosie's neck, wondering what to say to Jack in this unusual setting. Their usual pugilistic banter seemed out of place now that Jack had formally come calling.

"Charlotte, the dress becomes you. I hardly recognized my little business woman."

She glanced at him coyly. "I have other attributes than a head for business. You weren't really out training this mare were you?"

He started nervously, then agreed. "Why, me? No, not really. I really came to see you."

She turned her face up to him, eyes shining. "I'm glad you did. I wondered if you ever would come calling like a gentleman."

He took her arm, and they began to walk down the lane that was shaded by the trees. Eventually it would come to the road, but they never went that far. When they were past the view of the house, they stopped.

"Charlotte O'Neill you have turned me upside down, inside out. I sat in there a while ago and told your father I wanted you for a wife. Now what do you think of that?"

She stared up at him, the top of her auburn head just level with his chin. He was not smiling now, nor teasing. His eyes were black and bottomless, and his hair was glistening like coal in the moonlight.

"I . . . I don't know what to think. I've never considered getting married."

"Well, consider it. I want you."

"Jack, we'd better go back. I . . . I need time to think about it."

"What do you mean? Haven't I seen and talked with you every day for weeks now. You know me, Charly. And I know you. We are the same, one for one. You are the kind of woman I always wanted. Most women are too puny of spirit. There ain't nothing to them. They smile, they simper, they talk in quiet little tones. They are a washed out gray. But not you." He held her shoulders in a vise-like grip. "You're a flame, Charlotte, and you've been burning me up since the first day I saw you swimming in that meadow stream. You're like my wild horses that we round up now and again. They meet me head on, will for will, spirit for spirit. They're a challenge, and so are you. I couldn't love you if you weren't."

"Do you love me then?"

He considered that soberly for a moment. "I said it, didn't I? Yes, I guess I do, as much as I've ever loved anything. As much as I know how to love. I want you for my own, Charly, and I mean to have you."

She was watching his face. Was he the right one for her? What lay behind that darkly handsome face? She couldn't get her bearings, and she wished, for a moment, that she could kneel down to pray like Annie did when she was troubled. But prayer was only perfunctory with Charlotte, and she doubted that Annie's Mormon God would talk to her.

Jack was like a magnet pulling her, tearing her away from her comfortable, familiar world. In fact, he was physically pulling her, just off the road into the dark privacy of the trees. They stopped beneath a tall chestnut tree, and he backed her up against the trunk. The shadows of the trees hid them from the window of the house. Even Pat's sharp eyes could see no movement when Jack bent down and covered Charlotte's mouth in a kiss such as she had never had before. He seemed to consume her in his love, to draw her very soul inside him. A strange trembling began inside her, and she leaned against him, weak and frightened, somehow.

"Wait, Jack. Please wait . . . ," she whispered, struggling for control. "You must wait. I, I've never been kissed before . . . like that. It . . . confuses me. We've got to be sensible. Papa is growing to like you. I'm sure we can get his permission if we are careful. Oh Jack, oh Jack! I've never wanted to be married before. I've never been in love either, and I want it all to be just right. I don't want to do anything that would hurt Papa and Mama."

"What about me? Don't I count? Charly, damn it, I can't stand another day, another night without you. I'll be your prince. I'll sweep you up and carry you off on my horse. Come away with me tonight." He kissed the soft, shiny hair that was pillowing his cheek. "It's crazy. I don't even recognize myself. Never, not ever, have I ever done this, but I lay my heart out for you."

"I can't go away with you tonight. That would be the worst thing we could do. We would never get Papa's blessings."

She could feel coldness come over him. He stepped back and away from her. "All right, Charly, but don't keep me waiting too long. I ain't never set my heart out for a woman before. I just may not ever do it again." He turned away abruptly, starting back to the road, when a horse came trotting by, and Jack ducked back into the shadows with Charlotte.

"What's wrong?" she whispered.

"Nothing, only I don't want the whole town to know about us. I don't like gossip."

They stayed in the shadows, watching a man talk to Pat in the yellow light of the doorway. He kept gesturing to Jack's mare and Pat shook his head several times and then finally nodded. After a few minutes, the man rode back down the lane, passing the couple hidden amongst the trees. A minute later Pat came looking for them, and they met him only a few feet down the lane.

"It's late, Boughtman. We'll be saying good night." Pat's manner was short and strained.

"I'll be getting on then," Jack said. "Thank you for the walk about, Miss O'Neill." He climbed on his horse and started off. Tipping his hat to

Charlotte, he said, "Think on what I told you." Then he was gone, at a dead gallop.

Pat took her inside and told her to sit down. Then he told her and Margaret that the man who had stopped briefly was looking for a tall dark man on a bay mare. He was one of a mob of men who had burned down McGuire's farm, a place some six miles away. McGuire's elder boy had been shot, and had stumbled backward into the fire where he was burned to death.

"I'm afraid it's Boughtman," Pat said.

"Why do you say that?" Charlotte jumped to his defense. "Lots of men are dark-haired, and his mare isn't a bay. She's really a roan."

"Boughtman rode in here at a dead run, his horse all lathered up, and him as nervous as a cat. Looking all around every second, he was. Besides, there's one of the mob that was caught, and he confessed. Said it was Jack Boughtman leading them."

"I don't believe it!"

"Of course not! Ye're too durn pigheaded and stubborn to believe it. Ye've got yere cap set for him. It's plain to see. I saw the way ye looked at him, and he as good as told me he means to take ye for a wife. Open yere eyes, Charlotte. The man has been hanging around for weeks, and now that I think back on it, the only thing he wants to talk about are the things happening in Nauvoo. He pumps me all the time, and I, like a fool, have blabbed me big Irish mouth to him, telling him everything he wants to know. I would na' doubt that he was in on that apostate newspaper."

"Oh, Papa, you don't have any proof at all for these things."

"The word has been around about Jack Boughtman, and I've chosen not to pay attention."

"A bunch of gossip, that's all. Just because he's not Mormon like the rest of you, makes you all good and him all bad."

"Don't play the fool, girl. There's a mighty lot of evidence agin him. It wasn't until he took a fancy to you that he started acting civil to me and hanging around all the time. The man is in with the mob, and like as not, he's been using us. You as well as me."

She jumped up, smaller than he, and delicately made. Nevertheless, when her anger welled up, she seemed to grow in strength. "I do na' believe it. Ye can believe what ye want to. I will believe only the proof. A man is not proved guilty in this country by suspicion and rumor only."

"Charlotte," Pat said, as calmly as he could. "Jack Boughtman is not to come here again, and you are not to meet him again on the road to town."

"And pray, do ye think I've sought him out? He is the one who comes to me whilst I walk into town."

"Well, if he comes again, ye are not to ride with him. Do ye understand that?"

"I understand ye are mistrusting me again, and I do na' like it. And I do na' like being shouted at, like I always have been since I was a wee lass. I will ride with him if I choose, and see him when I choose, and if I choose I'll marry him, too."

"By heaven, ye will not!" Patrick shouted. His ruddy skin was brightly pink now, right into his hairline. "Ye are not such a grown-up woman, me girl. Ye are still my child and my responsibility, and I'll not let ye get mixed up with a man like Boughtman. My word, girl, do ye not understand? He is responsible for the murder of Billy McGuire. He's a murderer!"

"He is not! He is not!" she shouted back. "He's the first man I've ever loved or wanted, and ye just do na' want me to be happy."

"Land, child, that is all I want for ye. Can ye na' see that I love ye, and that is why I must protect ye from him?"

"Protect me from him? Ye can na' protect me from love all me life."

Margaret stood up, and moved between them. She put her hands on her daughter's shoulders and made Charlotte meet her eyes. "Dear, your father will turn you over to the care of a good man someday whom you will love and he will respect. He will not keep you from love and marriage. But surely, you must see that Boughtman is at least under grave suspicion, and you should hold back until the rumors are either proven or dropped."

Charlotte stood facing her mother, but challenging her father with emerald eyes flashing. Patrick calmed down first, and said gently to her, "I want only what is best for you—a man to love you and protect you. Be careful of your heart. Hearts are often wrong."

"Yes, sometimes. But so are parents." Charlotte stood her ground, glaring at him, still breathing heavily from her anger.

Pat started to flare again, but Margaret took his arm. He passed his palm over his forehead and sighed deeply. Margaret spoke for him. "Only promise us, dear, that you will not be alone with Jack Boughtman until we are satisfied of his innocence."

Charlotte was stubbornly silent. Margaret said again, "Promise us, dear. It is not asking too much."

Finally, hesitantly, Charlotte answered, "All right, mother. I'll not see him alone."

As it turned out, she didn't have any trouble keeping her promise. Jack didn't come into Nauvoo for a week, nor did he appear by the bend in the road as he usually had the past few weeks. Orin came to call every day at the store. Once he tried to take Jack's place in giving her a ride into town. She knew very well that her father had been encouraging Orin to court her, and she grew to detest seeing his hopeful face appear in the doorway of the store. Finally she grew sharp with him.

"Orin, have you nothing else to occupy you than coming here ten times day to see me?"

"I . . . well, I . . . hoped that . . . perhaps . . . "

"You hoped that with Jack out of the way I would fall in love with you, and we could be married next week. If I had been going to marry a schoolteacher, I could have done it a year ago. Nothing has changed between you and me. Now, please, give me some peace! I just want to be left alone to do my work."

"I'm sorry I annoyed you, Charlotte. I wouldn't do that for the world. May I come see you next Saturday?"

"No!" She was immediately sorry for her sharpness. It wasn't Orin's fault he wasn't Jack. It was just his weak manner. It brought out a strange anger in her. "Well, if you want, and if you will leave me alone til then. I feel like I'm suffocating."

He left, encouraged with the promise of Saturday, but inwardly fearing that she had slipped away from him.

Pat came out of the back office where he had been, discreetly leaving them alone together. "Ye are too sharp with him. He loves ye, lass, and ye'll have a hard time finding one better."

"If he's the best, then I don't want any at all."

Patrick stood looking at her, trying to understand, and trying to hold his temper as Margaret had pleaded with him to do.

"If that Jack Boughtman hadn't come into it, ye would have been glad of a gentleman's love. He's turned yere head with his smart-aleck, fancy ways, until common courtesy and respect ain't enough for ye."

Charlotte started to reply, but they heard a noise and, turning around, discovered Joseph Smith and Amasa Lyman standing in the doorway. They stepped into the room.

"We didn't mean to listen in," Smith said, approaching them. "Still, we heard you discussing Jack Boughtman as we started in." He bowed slightly to Charlotte. "Good day, Miss O'Neill. Your father and others tell me you are not only a lively dancer, you are also a business woman. You must have a good head on your shoulders."

Charlotte inclined her head slightly, acknowledging the compliment.

"A lady with your common sense surely has the wit to see past the front that Boughtman puts up. He has been plotting with William Law and others to repeal our charter and discredit Nauvoo as a respectable city. Moreover, we have substantial proof that Boughtman was the leader of the mob that burned out McGuire and killed his son; and now, last night, another farm was burned— Brother Thompson's place to the south of us. Luckily our boys were on the lookout and got word there might be trouble. We got there before the place was destroyed or anyone hurt. The man leading the gang was a tall man, bigger than anyone else, and he rode a palomino pony. I've seen the very one they described in Boughtman's corral."

Charlotte listened to him, looking him straight in the eye. When he had finished, she said pointedly, "Does he not sell his horses, sir? He sold you two as I recall."

"No! I settled for two after he had sold the horse I had bought from him to someone else."

Lyman spoke up for the first time. "Joseph is not the first man he has tried to cheat. There have been so many incidents of going back on his word— selling lame horses, selling sick horses—that our people refuse to deal with him any longer."

"I see. Well, you certainly have him convicted, don't you, gentlemen!"

Joseph spoke gently, "We do not convict anyone, Miss. That is for the Lord to do. We do, however, try to beware of the rattlesnake's bite." He paused a moment looking at her intently, perceiving the repressed anger and rebellion within her. She returned him gaze for gaze, and he privately thought she was a strong spirit, in need of much direction. As Lyman looked over the goods displayed on the counters and shelves, Joseph continued to search himself for the words that would open her eyes, or at least warn her. Patrick O'Neill was a favorite with him, as with the other townspeople, and he wanted to help him with this girl who had reputedly been a thorn in his side.

"You don't like me very much do you?" he asked her directly.

She was taken aback, but not embarrassed. "I don't know you, sir. I know what other people tell me about you, but *I don't judge* by gossip or other people's opinions. Mostly I don't like some of the things you advocate and are forcing on the people because of your position, such things as plural marriage."

Again he smiled, teasing her gently. "I fail to see how you can possibly be concerned when you have yet to be married at all."

She grew cold at his teasing. "When I marry, sir, it will be to a man who will love and cherish me alone, not send his wandering eye to every new face he can find. It is a pity that one woman cannot fill your needs, but I intend to fill all of my husband's."

His face was impassive. "Young lady, I fear you are the one judging on gossip, for you haven't any facts at hand at all on my relationships. That is between Emma and myself."

Then his blue eyes began to penetrate, and he took her arm in his hand, leaning nearer to her. He spoke with a vibrant intensity. "Charlotte O'Neill, the day will come when you will be humbled sufficiently to accept any law that God reveals. You will find that plural marriage is much preferable to the godless, loveless, immoral practices of many people outside this church, and you will weep in sorrow for your lost chance to be even a tenth wife to a good, kind man. If you accept Jack Boughtman, you reject God. The two are mutually exclusive, for Jack has put himself at opposition to God."

Their faces were inches apart, and his words were for her ears alone. She saw something come over him that amazed her. It wasn't only his eyes that conveyed the importance of his words. It was a burning in her head and in her heart, and an energy seemed to come from him, passing into her. She didn't know what to say. She was groping.

"President Smith . . . what should I do? I . . . ?"

"Leave him alone."

"But there is much between us."

He glanced at her sharply. "How much?"

She blushed. "Not anything . . . immoral. He has asked me to marry him and claims to love me."

"Do you love him?"

"I don't know. I'm . . . I'm very confused. I don't know what to think. When I'm with him I feel as though I love him, but you are shaking my confidence."

"When love comes from God, it makes you a better person, and it is certain and secure. If a man is honorable, he will honor your womanhood and treat you with respect. He will build you up to be a queen, not degrade you. You are a smart girl, but you need humility and teachability. Before the Lord can use you or exalt you, you will have to develop those things."

She was held by his eyes, and could not look away. "The last time we met," he said, "You had not decided how to choose. Have you chosen yet?"

She knew what he referred to. Was it really only a few weeks ago she had met him at the Church social? It seemed like an eternity. Charlotte blushed when she recalled their conversation and her open challenge of him as a leader. He had told her to choose soon whom she would follow, and she was farther away than ever from making a decision.

"No sir. Sometimes I think I believe . . . . But then again . . . well, I just don't know."

Joseph's face was filled with sympathy and concern. "Be careful how you choose, Miss O'Neill. It is very important. We may not know until the last heartbeat of the universe what effect one single choice has upon our lives, but it is true that the sum of all our choices will determine our exaltation or damnation."

He paused, then said very distinctly, "Life is always a challenge and a choice. And beware the wrong choice."

She was still thinking about it all as she lay in bed that night beside Annie. Long after Annie had kissed her cheek, said good night, and turned over, breathing deeply in her untroubled sleep, Charlotte still lay awake. In the quiet of her own thoughts, with no one to scold or preach, she could admit to herself that the Mormon prophet was an unusual man. She could like him. In fact, she did like him, and not even reluctantly. She smiled to herself. Orin would be so pleased. Not that she could tell him, for it would give him unfounded hope. A nagging, intrusive thought began pestering her reflections. Over and over, the impression came that she must pray. Over and over, she brushed it away. Soon it became more difficult to lie quietly in bed. The thought would return, and she would tense with the battle inside her. At last, she whispered quietly into the darkness, "Oh all right. What harm can it do?" Charlotte slipped over the edge of the bed and onto her knees.

Now, what am I to say, she wondered as she knelt. She felt rather awkward, for prayer had always been perfunctory with her, an indulgence of Patrick's insistence.

"I don't know what to say." Her lips moved in soundless conversation with an unseen God. "Why am I down here?" she asked. "Am I supposed to talk to you about Jack? Why can't you talk to me in bed?"

Then came a deeply peaceful feeling, lowering upon her head and shoulders like a warm blanket. A catch came to Charlotte's throat and her heart began to throb. The darkness ceased to be empty for her, and a profound sense of communion with another soul filled the room.

"Oh dear, dear God, tell me what to do. I don't trust my Papa's commands, and I'm afraid to trust Joseph's. I want to love Jack. Surely he is not the monster they make him out to be. You tell me what to do, and I'll do it."

Then came an impression of gentle love, and, with it, a dim image of a person she had never known. She seemed to sense a man, radiating strength and kindness, and an ocean of love rolled in, engulfing her. Clear as the evening bell, she seemed to hear the words, "This is love." She put her head down on her hands folded on the mattress, and she wept with the overflow of an emotion she had never experienced so fully before. It was so tender, so sweet. Hurt, yes it hurt, like a muscle unused to stretching.

She knelt there for a long time, auburn hair burnishing white nightgown and bed sheets. Her heart was at peace, her passionate nature calm and at rest. For once, all strife seemed to have vanished, and love had settled about her shoulders like a warm shawl. She had no will to move. She would like to have stayed there forever, but her communion was disturbed by soft noises outside. Footsteps! Charlotte looked up, absolutely still, listening to the footsteps. Then she crept cautiously over to the window, visions of torches and men on horses filling her mind. At first she saw no one, but the sound of footsteps was becoming clearer. She drew back so as not to be seen. They stopped outside her window, and she could see a shadow on the far wall of the room. It was Jack. Her heart turned over. Yes, it was him, standing still, looking intently into the dark room. She moved out of the shadows, her face inches from his. He was startled but reacted quickly.

"Charly, I'm glad I've found you. Come out here where we can talk."

"What is it? Are you in trouble?"

"Hell, no, I just don't want your whole family to wake up."

"All right, walk down the lane a bit. I'll be right there." She grabbed her shawl, wrapped it around her shoulders, and thrust her feet into her shoes. She glanced at Annie. Her sister was still sleeping soundly. Forgetting her promise to her mother, she quietly climbed out of the window. Once out, she found Jack in moments. He was waiting at the head of the lane and opened his arms when she came hurrying across the grass.

"Charly, I've been so lonesome for you. I never thought I'd say that."

"I've been lonely without you, too. Where have you been?"

"Tending to my own business. Your Pa sent word that you didn't want to see me, and for me to stay away."

"He did! I never said such a thing. When you left last Saturday night, we had an argument about you. He said you were in on the burning of McGuire's place and that you killed Billy." She watched his face, hoping to find assurance there that Patrick and Joseph were wrong.

He held her shoulders with his great hands. "You didn't believe him did you?"

"Well, no, but since then, there has been a lot of proof against you."

"You can't believe I'd do such a thing. I may be rough and wild in my ways sometimes, but I'd never burn a farm or kill a little boy. There are lots of people in town who'd like to blame me for the persecution because they don't know who it really is."

She didn't answer, but watched his face, searching it for the truth.

"Charly, last week I told you I wanted to marry you. This week has been hell without you. I can't even run the ranch, for mooning around like a schoolteacher after you. I can't stand it. I'm going crazy. Come away with me tonight and marry me. We'll do it good and proper. There's a preacher in Montrose that says a fine sermon. He'll marry us pretty as you please."

"I don't know, Jack. Papa would have a fit."

"Well, let him. You can't stay a little girl all your life and do just what your Pa wants. You've got to make up your own mind, live your own life. Live it with me, Charly. I love you." He held her easily in his arms, kissing her as gently as he could, murmuring his love words against her lips.

"Please, Jack. I'm so confused. I don't know what I want. But I don't want to hurt Papa and Mama. I can make him be reasonable, I think, if we just wait 'til all those rumors are over."

Patrick's voice broke into their secret world. "Ye'll never have me blessing with the like's o' him."

"Papa!"

"Ye'll sneak out to meet him! What else will ye do, Charlotte O'Neill? Ye broke yere solemn promise to me. Ye vowed ye'd not see him alone, and here ye are, in his arms, kissing him. I should have broken yere rebellious ways when ye were a child. I were too easy on ye."

"Papa, I'm sorry I broke my promise. I really didn't even remember it when Jack asked to speak to me. He came to ask me to marry him."

The rage in Patrick broke open then, and he roared, "Oh, he did, did he? And ye, ye foolish girl, no doubt ye were about to run away with him. If ye bed with him, ye bed with the devil, and I'll not have a son of Satan in me family. Jack Boughtman, if ever I see ye again on my land, I'll shoot ye without asking a question first. Charlotte, git into the house. Boughtman, get gone, or do I have to shoot ye where ye stand?"

Charlotte ran to him, threw her arms around him, pinning his arms to his side. She pleaded, "Oh Papa, don't do this. If you send Jack away, you send me, too."

"What? Will ye defy me?" Patrick bellowed. "And for the likes 'o him! Have ye no sense in yere head? Ye know all that I know about him. He is a murderer, he is a cheat, and a liar. The prophet hisself warned ye against him. Be smart, girl. Be smart for once, and obey yere father. Git in the house now."

"No, I won't! I'll not be treated as a child any longer. Ye are the pigheaded one, Patrick O'Neill, or ye would see I want to keep yere good will. I do na' want to hurt ye or make ye mad, but ye persist. Well, ye can na' order me around. I am a grown-up woman, and if I want to marry Jack, I will— without yere blessing if ye are determined."

"No ye won't! I'll tie ye to the bedpost first!"

Charlotte was aghast, and then furious. "Oh, ye will! We'll see what we shall see. Ye'll never be able to hold me. I am not a slave, for all ye try to make me one. Jack, where's yere horse?"

Patrick stood with his red head lowered, legs planted wide apart like a furious bull. "Do na' come back Charlotte O'Neill, if ye go with him."

Margaret came running from the house, having heard the foolish words of her husband and her daughter. Annie was close behind, and cried out, "Charlotte wait!"

Margaret ran after her willful daughter as she and Jack were making for the horse. "Charlotte, child, wait! Please don't do this. Wait, dear, wait. Let's talk it out. We can talk more calmly in the morning."

Charlotte turned on her icily. "Mother, I can na' talk with him. And I'll not risk being tied to the bedpost like some animal. I tried to keep me temper. I tried to reason with him, but he'll never change. I'll go with Jack now, but I'll see ye again soon Mother. You, too, Annie dear." She looked beyond them to her father, and spat out the words, "But may I never see his face again."

"You don't mean that, Charlotte. Your father loves you. Don't go away like this."

Annie held onto her leg as she sat behind Jack on the horse. Annie's little heart-shaped face was white as her nightgown in the moonlight, and perfectly masked in anguish. "O Charlotte, don't leave, please don't leave! I'll never see you again, I just know it. And I love you so. I'll just die without you."

"I'll see you, pet. We'll not lose one another. I have to go. I can't live here anymore and be treated so by him. I'll send for my things in a day or two."

Patrick's anger had crumbled as he saw his daughter slipping away from him for good. Remorse humbled him, and he started slowly toward them, "Charlotte, lass, will ye forgive me, I . . . ."

She turned on him in fury as the horse danced a circle in its impatience. "No, I will na'. Ye have whipped me for the last time. The switch was never as bad as yere words and belittling. Ye have tried to love me, and I have tried to love you, but ye have no faith in me, no trust. Ye think the worst and never the best. No, by heaven, I'll not forgive ye again. And I'll not go through it again. May God bless me to never see yere face more."

They rode off, Jack triumphant, having let the two Irish tempers run their courses, and Charlotte angry, frightened, clinging to him with her nightgown fluttering in the breeze. Patrick stood with bowed head, the tears coursing down his cheeks and gruff, animal sound breaking from his heart. Margaret went to her husband, covering him with her arms and long brown hair. Annie sat down alone on the ground, looking after Charlotte as long as she could see the gleam of white, crying for the person she loved most in the world.

CHAPTER 3

Charlotte slept, that night, curled up in a big rocking chair in front of the cold, dead fireplace in Jack's tiny bachelor's cabin. When they rode into the ranch she had cooled down but was drained by the emotion of the evening. She could hardly even feign interest in the horse ranch that was to be her new home. Jack's cabin was dark and smelled of stale cigar smoke and old whiskey bottles stashed in odd corners. There was only one tiny hole cut for a window, one bedroom, and one all-purpose living area with a large open fireplace for cooking.

Jack was jubilant, and proudly talked about his "place". Charlotte felt like crying. She was tired, she was lonely already, and she realized that she had started down an unknown road. But she was frozen into solid ice inside, so no tears came.

He led her into his hole of a bedroom and said, "This is your bed now."

She turned to him, her voice flat and final, "Not yet, it isn't. When we are properly married, like you promised, before a minister, and I say 'I do,' then I'll make your bed mine. Not til then."

"Aww come on, Charly. We'll go tomorrow to the preacher. Ain't nothing wrong with getting started now."

She turned around and started out the cabin door. "I'll sleep in the barn."

"And won't you look purty when the hired hands find you in the morning? You in yer nightgown, with straw in yer hair! Oh, come on back in here. If you're so anxious to protect your virginity one more night, you can sleep out here by the fire. As for me, I'm sleeping in the bed."

Reluctantly she realized that he was right about the hired hands. Instinctively she knew that she must start her life as Jack's wife with dignity before the working men. She was terribly tired; even her bones felt weak. It was tempting to crawl into Jack's arms and let him take her to bed. Still, something held her back. She knew him well enough to be sure he would take something for nothing as long as he could, and she would never again have the drawing card she now held if she gave into him. If she was to leave her family and home, she would be certain to exchange it for something equally secure. Besides, she wasn't ready for him. There would be no joy at all for her tonight in giving herself.

She sank down in the rocker. "I'll sleep in the chair." Her face became a little girl's, and she spoke wistfully, "Will you make me a fire, Jack?"

"You cold?" he asked surprised.

"My bones are."

"Well, now I don't know. I ought not to make it so easy for you to sleep out here. My bed will be warm." But despite his disappointment, he couldn't say no to the pitiful, small figure before him. "All right, Charly. Just for you."

From the corner he hauled out fire logs and soon had a fire started. Charlotte had curled up in the chair, auburn hair tumbling over shoulders and arms. Her freckles stood out in stark contrast to an otherwise white, pinched face. When Jack turned around to her, the fire flaring up, her eyes were closed and her head pillowed on her arms, knees drawn up to her chest. What was she hugging so tightly—her girlhood, her dreams, her tears? Jack looked down at her and grinned.

"You're mine, Charlotte O'Neill Boughtman," he said. "And we are gonna have some fine time together, little spitfire." He thought of the wild horses he loved so well, and how he loved to tame them, to make them submit. Well, he would master her too, but he was glad she put up a good fight. It made the victory sweeter. He bent over her and kissed her hair and left her to her dreams.

Jack sent Jimmy John over to the O'Neill place the next morning to get Charlotte's things. He took the buckboard and came home with it loaded, not only with clothes, but wedding gifts as well—samplers, bowls, a tablecloth Annie had cross-stitched for a year, a shawl, and a hooked rug that Charlotte had made before she started working at the store. There was a bolt of blue cloth Patrick sent, and some thread and pearl buttons. She put them away, and it was years before she touched them again. Her mother sent the cedar chest she and Annie had been storing their trousseaus in for years. In the trunk were nightgowns, linen, quilts, underclothing, and a lilac sachet that filled the dark bachelor's house with springtime smells and sunshine. That was when she cried. Sitting there with all those loving gifts about her, her nightgown on, and the sleep still unwashed from her body, she cried. She would have loved to run home and hug her mother and sister. She would like to build her house just down the road from them, and walk back and forth every day to visit and share her new life with them. She wished John Patrick were not so far away, but he was in New England on a mission.

Oh, she wanted the sun to shine! Today was her wedding day. She wanted her heart to sing! That's the way it should be on your wedding day. She wanted so desperately to be happy. She would be happy! She'd show them all how glad she was that she had left with Jack. She was so sure that, when it was all settled, she would be happy she was Jack's wife. She impatiently brushed away the tears. But now, she had to be about her day. The first thing to do was get into some decent clothes so she could get out of this dismal, dirty cabin and out into the bright June day.

Jack had brought her a bucket of water, apologizing that it was not as good as a stream in the meadow. So, barring the door with a chair, she quickly sponged herself off and climbed into one of her dresses. Hair neatly coiled into a braid at the nape of her neck, she sprinkled some of the lilac sachet into her bodice and pinched her cheeks. Then she went to survey her new kingdom.

In the sunshine it was far more inviting than the big, dark shadows last night had promised. Whereas Jack's own living quarters were small—as reflected his needs—the barn was immense, containing twenty stalls for horses, a large area for parking his fine buggy, and one huge wall where saddles, bridles, bits, whips, and blankets were hung. Overhead was the hayloft, dripping bits of hay from between the rough boards. And it all smelled marvelous to Charlotte, the horses being curried, the sweaty odors of working horses, the fresh hay, the leather. It was almost as exhilarating as the meadow with its prairie flowers and fresh water.

Detached from the barn and removed about fifty feet, was a shed with bellows and a blacksmith shop. Across from the barn was a bunkhouse where four men slept. It was as rough and dank as Jack's own cabin, with just enough room for four narrow cots, a couple of stools, and some pegs on the wall where clothes were carelessly slung.

She loved it all. She had known she would, adoring horses as she did. There must have been better than two dozen animals between the corral, the barn, and the pasture. Rose O Sharon was there in the pasture and came to Charlotte quickly, as she stood on the lower beam of the fence and whistled. Charlotte rubbed the long, sleek nose, scratched between her fine, velvet soft ears, and looked the horse over carefully, unwillingly remembering the description of the mob leader who had ridden a bay mare.

Jack was at the corral, training a colt. She watched him from the pasture fence. The colt was a paint, perhaps a year old, and Jack was putting him through his paces around the ring, first trotting, then cantering, then walking. After a few minutes he led the horse over to her.

"I want to learn to do that, Jack."

"What, go through your paces? I'll see that you learn them pretty quick." He was laughing at her, his black eyes shining.

"You know what I mean. I want to train horses, too. I could do that. Horses will do anything for me."

"A woman! Women don't train horses."

"I do! Or will, at any rate. I want to learn. Please! I won't be a bother. I'll just watch for a while, and you could let me do the easy things first, and teach me the harder part later."

He cocked his head, tugging at the brim of his hat, and squinted. "I'd be the joke of the state if I let my wife train horses. No self-respecting citizen would invite either of us to their home. And no one would buy an animal you trained, anyway. No, you're welcome to watch and to help birth the foals, but training and breaking is a man's work."

"Oh, fiddle. You don't really care what people think. You never have before." Now it was her turn to tease him, "Is marriage making a respectable man of you?"

"Hell, no! Nothing will. But I don't like to be the joke."

"Who'd have the nerve to laugh at you? There isn't a man around who would. You're your own man and everybody knows it. Well, I'm my own person, too. Not right away, but soon, Jack—just for fun—maybe I could do the easy things, and nobody would notice."

He put his arm around her and squeezed. "Oh, all right, Charly. You can be a horse trainer. But only after you're my wife." He tipped her face up to his and grinned in pure pleasure at her fresh, clear face and sparkling gray-green eyes. "When are you gonna be ready to face the preacher?"

"Whenever you are."

"Soon as I'm finished with this colt, I'll go and put on my Sunday clothes for you. I've even got some of that pretty smelling water I'll splash on. What'll you wear for a wedding dress?"

"I . . . uh.. well, I hadn't thought about it. I dunno . . . "

"That yeller dress! Wear that yeller dress. You're as pretty as a picture in it." He slapped her bottom. "Go get your ribbons on, Charly, we're going to a wedding!" He turned and yelled at his foreman. "Emery, go get cleaned up, and slick down your hair. You're gonna be a best man. The best you've ever been, I reckon."

And so it was that Charlotte O'Neill was married to Jack Boughtman; she in her yellow Sunday dress with the hopeful lace and ribbons, and he in his one good suit, with a sprig of lilac sticking out of his button hole. Jack took the rest of the day off from ranch work and drove her out to a meadow, almost as pretty as the one behind the O'Neill home place. In the meadow, on the quilt that her mother had made, he loved her, and all that she remembered afterward was the birds chirping overhead. The birds and the glorious spring sunshine, through the leaves of the trees, were the only happy things she had to remember. Jack was neither tender nor gentle. He loved her, so he said. Charlotte wasn't sure why his love left her so cold and empty. It wasn't that she wanted a lot of pretty words or flowers. Jack would never be like that. But she wanted to give herself, to be allowed to love, to really love for once. There was a river of emotion within her, as wide as the Mississippi and just as tumultuous. She had often felt the fullness of emotion, sometimes manifested as anger, sometimes as pure joy, and she had always imagined that when she loved, the great river would engulf her with a happiness that she had never known. But when the afternoon with Jack was done, the love that had been building inside during the last years was all dissipated, like water disappeared into a burning sand, and she felt more empty than ever before.

They rode in Jack's carriage through the countryside for a while, and when they went back to the cabin, the sun was shedding its final glimmers over the sky. Jack was talkative, bright and strutting like a peacock. Charlotte was unusually quiet. Inside, the cabin had been cleaned by Lily, Emery Jordan's quiet, plain wife. Charlotte felt a rush of thankfulness to a stranger who had troubled to scrub the rough floor, to clean it up and light the candles. Her rug was on the floor, her tablecloth on the table. Her crockery bowls sat on the hearth, and her shawl was laid over the back of the rocking chair.

"Well, Lily did a good job. I asked her to clean the place up a bit, and she laid it out right fancy. Even fixed us a bite to eat. Smells good."

"Jack Boughtman, you just wait til we sit down to the table proper and say grace, before you go dipping into the pot."

"Just got me a wife for a few hours, and she starts telling me what to do." Jack pierced her with his black stare. "Charly, didn't you learn nothing from this afternoon?"

"I learned a few things," she said sarcastically. "What did you have in mind?"

"I'm the master here," he said quietly.

"Oh, really! Well no, I can't rightly say that was what I got from it all. Mostly, I learned its not my love you're wanting, it's a roll in the grass."

"Same thing, little miss, same thing."

"No, it isn't."

He frowned. "What more is there?"

She sighed, "I don't really know. I've never had it, either, but I know there must be more." She sat back wearily in her chair, the dim light glinting gold on the copper of her hair. Jack grinned at her, pleased at his new possession, then shoved back his chair. Moving smoothly and quickly as a cougar, he drew her up from her chair into his arms.

"Take your hair down, lady. I like you with your hair down. Don't you know I love you? Nobody could love you like I do. Love ain't a lot of flowery words like that schoolteacher used to mumble. Love is my own two strong arms, and your two warm lips. I love everything about you, Charly, your flashy eyes when you're mad, your square jaw when you play tug of war with me, your coppery hair falling down on my hands, and your softness in the meadow today. If there's more to love than what I had in the grass with you this afternoon, I sure can't imagine what it would be."

She didn't move, so he reached out and loosed the pins, letting her hair down. It fell almost to her waist.

He stepped back for a moment and looked at her. "That's the way I remember you beside that stream. You're a beautiful woman, Charlotte Boughtman. And you're my woman; don't ever forget it. My woman, and no other man had better look cross-eyed at you."

He scooped her up into his arms and covered her mouth with his. "Wait, she pleaded. "Wait, Jack, can't we enjoy the firelight together, or sit out under the stars for a while?"

He didn't answer. He had a wife, and he loved her and wanted her now. That was all he knew or cared. Like a butterfly in a cocoon, Charlotte sensed that somehow there was sunlight and warmth and glory in life. She had an image of real love imprinted in her heart, the tender soothing love she had bathed in momentarily only a few nights before beside her bed in the O'Neill home. Oh yes, Jack loved her in his way, but not the sweet, gentle way she longed for. So she was never able to give him the tender blossoms that struggled to bloom in her heart. She never gave him an inch, and that was just how he wanted it.

They had been married a week when Jack came bursting into the cabin one night about eleven o'clock, after having been to a Masonic meeting. He was elated over the newest piece of news that Joseph Smith had been arrested again and brought before the Governor of Illinois, Mr. Thomas Ford.

"This time we're gonna get him! He's been loose long enough, carrying on his own private government, organizing his own army, running for President, just as though he thought he was as good as Judge Douglas. We'll get him, Charly, we'll get him. The no-good, pantywaist Mormon prophet is locked up for good this time. He won't get away again."

Charlotte had been asleep, and was now trying to get her bearings and understand what Jack was raving about. She didn't comprehend exactly what was going on, but she did realize that Jack had a deep resentment against Joseph that she had never been allowed to see before.

"What happened? I don't understand. Is he in jail now?"

"Yes, in Carthage. He turned himself in this morning and went before the governor on charges of treason."

"Treason! Why treason?"

"The newspaper. The Nauvoo Expositor! He had it taken apart and won't allow freedom of the press. That's treason, Charly, because our Constitution guarantees freedom of the press."

"Yes, but that thing—did you read the lies it published?"

"What d'ya mean did I read it? Sure I read it. I helped finance it. Somebody had to speak up against King Smith over there across the river."

She began to be uncomfortable. So he was behind the paper. After President Smith had told her in the store that Jack was mixed up in it, she had read the one and only copy the paper had published. Even in her eyes, it was obviously false and inflammatory. Outsiders would not be able to spot the discrepancies in the allegations made, but those who knew the religion and the leaders certainly could. That memory brought back, like a ripple of clear water, the prophecy and warning Joseph had made to her. She sat up in bed watching Jack and listening to him. In a few minutes, he calmed down and noticed her silence.

"What's the matter? Anything the matter?"

"No, at least I hope not," she said looking down at the bed, smoothing the quilt Margaret had sent. Then she looked up and caught his eyes with her direct gaze. "I didn't know you were behind the Expositor. Papa said you were, but I didn't believe him. Is there anything else, Jack, that you have been in on that I didn't know about?"

He grew wary. "No, no there ain't nothing else. A fella can give money to a worthy cause can't he? I think Smith ought to be knocked down from his pedestal, and I'm glad to see it happening. But that's all. What else do you think? You think I was in with that mob, don't you?"

"I don't know what to think." She took his hand in hers and held on tightly. "Oh Jack, I don't want to believe that. You're not like that. I couldn't love you if you were. Sometimes you're rough and wild, but you're not mean. Just a little hot-headed, like me. That's what I told Papa, and I do believe it."

He patted her hand and then drew her into his arms, caressing her hair, patting its soft fullness. "That's right, Charly, I'm a wild one all right, but I ain't the burning type."

He kissed her then, and she began to forget the doubts. She always felt as though she were stumbling in a morning fog when it came to Jack, groping for some solid footing, afraid of unseen pits. She sensed there was a depth in him that she had not yet discovered and was reluctant to confront.

Two days later, on June 27th, Jack left the ranch in Emery's hands, saying that he'd be back by the next morning, if not before. He came in to say good-bye to Charlotte. She was rushing through the cabin, cleaning up, and starting afternoon dinner to simmer so she could go on to more pleasant things. Jack said he was headed south to arrange the buying of another five horses. He expected to be gone most of the day, and she was not to worry if he got back late. She followed him out of the house and waved good-bye. He had saddled the palomino, and his rifle was strapped to the saddle.

"What's the rifle for?" she asked.

"Shoot some rabbits, maybe." He answered with a smile.

She frowned. "What for? You hate rabbit stew. You told me so two nights ago when I made some for supper."

"All right then, I'll shoot some turtles. I like turtle soup."

"How about some snails? You could probably blow the shells right off with that, and I could make you snail custard pie."

"Oh, shut up! I'll take my damned rifle with me if I want to. Now quit asking so many questions. You're starting to act like a woman." He leaned down and kissed her soundly, then started down the road at a canter. She could hear him whistling as he went.

Emery came up to her side. He was a gentle man, about thirty-five, who looked like the horses he trained. He had a long face, a long thin nose, and a shock of wiry, brown hair that fell straight over his eyes. He had taken an immediate liking to the new missus, telling Charlotte all she wanted to know about training horses. He was, at first, a little shy with her, because, as he told his wife, Lily, she was so pretty it embarrassed him to look her in the eye. But he soon got over that. Charlotte was completely unpretentious. She moved among the hired hands asking millions of questions, bringing them lemonade or just plain, cool water when they got hot. She was part of something she loved for the first time in her life.

"He's in an all-fired hurry to git somewhere, ain't he?" Emery said.

"Yes, I guess he's anxious to see those new horses he's buying." She turned slightly. Her eyes were on a level with the foreman's. "Emery, does he ride that palomino very often?"

"Pretty good bit. He says it's the fastest horse he's got, though to my mind it ain't the steadiest. And I value a good, dependable horse any day over a fast un."

"He's been gone quite a lot since we were married. Does he always go out so much?"

"Well, he's never been one for staying home. Course there ain't never been nothing to keep him home til now. I 'spect he'll settle down right soon. He's been kicking up his heels for a right smart o' years. A man don't forgit his old ways right off." He patted her shoulder, and offered one of his many platitudes. "Just 'member now, missy, that everything comes to him that waits. Just 'member that now."

It was midnight when Jack came home. There was a curious air about him, agitated and excited, though he tried to hide it from her. She asked him about the horses, and he said they were fine animals. They would get them in a few days. She asked him several times if anything was wrong, and he answered twice that nothing was wrong. He wasn't acting any way at all. She was imagining it. The third time she brought it up, he slammed his fist down onto the table and said to shut up. She was picking, and he couldn't stand a woman that picked. She quietly dumped the plate of food she was preparing on the dirt outside the cabin door and told him to fix it himself, she was tired of picking and fixing. She went back to bed.

The next morning about nine o'clock, a rider came galloping in to tell them the news. The Mormon prophet, Joseph Smith, and his brother, Hyrum, were shot to death in Carthage Jail. Charlotte turned quickly toward Jack. He was politely sorry.

"Now ain't that awful. Never liked the man personally, everybody knows that. But it's a shame to see anybody shot down."

"What do you mean 'see'?" she asked in a measured voice.

Jack replied quickly, "Well, I didn't mean to actually see him." He turned to the boy spreading the news, "Did anyone see it happen?"

"Plenty of people. There was a regular mob that thronged the jail and five or six of them that was there shot him at the window of the jail with their rifles. He's dead all right, and Hyrum along with him. Guess that'll be the end of the Mormons, now their precious prophet is dead." He trotted off down the lane to tell the rest of the neighbors.

Charlotte stood staring after him, fixed to the spot. She could see a rapid succession of images darting through her mind—Joseph's eyes penetrating her soul when he said "Choose soon, you can't afford to wait much longer", and then later when he said she would someday accept any law that God revealed. She felt, in a sudden flash, the realization that she did like the man, more than she would have admitted to her family, and the news of his death came as a loss to her. She could see Joseph's blue eyes, intense and concerned when he said, "Beware the wrong choice." And she saw Annie's pinched little face gleaming white in the moonlight, as she rode off in the dark with Jack.

With great effort she turned to her husband. But he had already begun to walk off toward the corral. Charlotte caught up with him and grabbed at his arm. "Were you there?"

"Hell, no, I wasn't there! You know where I was, away up north buying us five new horses."

"Jack Boughtman, don't you lie to me. I may take a lot from you, but don't you ever lie to me."

He returned her intent, searching look with a steady, reassuring gaze. "I never would." Then he slapped the pigtail braided down her back. "Come on, let's go watch 'em break that sorrel."

Nauvoo was in mourning. Patrick had closed his store. The Quorum of the Twelve Apostles had advised everyone to lay low and keep to themselves, and above all, not to meet in large numbers so that they could be accused of rioting. They were following those instructions to the letter. It was not hard to do. Everyone was wrapped up in his own private grief at losing Joseph, the beloved prophet, and his brother, Hyrum, pillar of the church.

Charlotte sent two notes of condolence, one to her parents and one to Orin.

Dear Mother, (she wrote)
 I just heard about President Smith and Hyrum. I know how much you admired Joseph and how grieved you all must be. I can only say I'm sorry that it all ended so. I never believed that

anything or anyone could kill him. He seemed so much bigger than life. You know I never believed in him like you do, but a few weeks ago I talked with him for a few minutes and finally understand why you think so much of him.

Thank you for the gifts. They are very welcome, as our cabin is really a man's cabin. Jack has promised me a brand new house on some land that he has near the river, a few miles north of here. I am anxious to get started on it.

Jack is good to me, and I am enjoying the horses especially. I have his promise that I can choose any horse I want for my own. But I don't want one that someone else has trained. I want to train my own, so that he's mine and no one else's.

Please don't worry about me. I wish I could come see you. Perhaps we can meet someday in the meadow and have a picnic. You are always welcome here.

Affectionately,

Charlotte

She sat for a long time over her second letter. Finally she began.

Dear Orin,

What can I say? My feelings are very keen, and I hardly know how to tell you how sorry I am that President Smith has been killed. I know what reverence you had for him, and I'm sure you are suffering greatly now. I wish I could say something that would help, but I know that I cannot help you now.

Always I teased you and treated you awfully, but I want you to know that in my own way, I respect you greatly. I only wish that I loved you. I am afraid, however, that my love and my hand in marriage belongs to another. I am Mrs. Jack Boughtman now, as you must already know. I am happy, and I hope that you will someday love another and find happiness yourself.

Sincerely,

Charlotte O'Neill Boughtman

The next day little Joshua Logan, the twelve-year-old farm boy that lived three miles away from her parents, came riding up and delivered a letter to her. When she asked if it were from her father, he answered "no", blushed and said that Annie had written it and given it to him to deliver.

He left quickly, glancing around continually, as though expecting the devil himself. Charlotte tore the letter open and consumed the contents voraciously, as though it were the first meal she had had in months. Annie had written only a short note, asking her to come for a visit. She suggested that they could meet in the meadow. Her penmanship was neat and tidy, like her samplers, and Charlotte felt a choking lump deep in her throat, as she allowed herself to visualize her sister. Annie said she would be waiting tomorrow at noon in their favorite spot and would wait as long as she could.

Jack was away the next morning, visiting in town. Charlotte was glad. Not that he could have made her miss the meeting, but he could have spoiled the morning by quarreling with her. At eleven o'clock, Emery saddled up Rose O Sharon for her, and she started off to Nauvoo. The town was deserted as she passed through it. She passed Pat's store and saw the door was bolted. That meant he was at home, and she was disappointed, knowing that she could not visit her mother. The ride was leisurely and pleasant, the day warm and languid. Actually Charlotte had seldom been so happy. She had her own home, humble though it was, and her own man, and plenty of time to spend with the horses she had come to adore. So she whistled as she rode, happily knowing she was very unlady-like.

As she neared the trees that ringed the meadow, she clucked to Rosey and off they went at a trot, slipping easily between the trees, then breaking out into the sunlit meadow. Annie was there as she had promised. Charlotte threw the reins down and jumped off the horse, rushing to the younger girl and throwing her arms around her.

"Oh Charlotte, I was so afraid he wouldn't let you come."

"Don't be silly. Of course, I came, pet. I wouldn't leave you here, waiting all alone. I've missed you so. I got the things you sent and they were lovely. I have all of them set about the cabin, and it seems almost like home. The samplers are hung up on the fireplace, and the bowls are all lined up right on the edge of the hearth. All the trousseau things are right at the foot of my bed, just as they used to be at the foot of our little bed at home. I take them out almost every day and look at them and touch them, and I can see you so plain, Annie. You don't know how many times I've remembered the things we've always done together and the games we played in the woods. It seems like a hundred years ago. Annie, oh Annie, come home with me and live with us. I miss you so."

Annie answered carefully. "Well, maybe I could pay a visit if you're sure 'he' wouldn't mind."

"Who is 'he'? Papa or Jack?"

Annie looked down, embarrassed. "Oh, Jack," Charlotte said, as she understood. "Jack is what's bothering you. Listen, pet, Jack's all right. He's a little rough in his ways, and he teases a lot, but he's all right. Really he is. He loves me a lot, you know." She paused for a moment as she thought about her marriage. "And he is very good to me. He bought me this riding outfit. Isn't it pretty? He's going to build me a house, a fine house with an upstairs and windows and a flower garden, just like the grand houses in Nauvoo. We've already been pacing off the foundation of it. Jack really won't eat you up. Why, I'm happier than I've ever been before."

"You look happy enough," Annie said a little hesitantly.

"Of course I am," Charlotte said. "What made you think I wouldn't be?"

Annie blushed and fumbled for words. "Just the things we've all heard. I mean . . . well . . . they say . . ."

"What do they say? What have you heard?" Charlotte demanded.

It was painful for Annie. She didn't want to make her sister angry and spoil their afternoon. "Nothing much, just rumors, I guess."

But Charlotte wouldn't let it go. "What rumors? Have people been talking about me?"

"Oh no, not you."

"Jack, then!"

Annie didn't reply.

"Yes, it is Jack, isn't it? What do they say? Tell me Annie. I want to know." Her eyes were growing green, and Annie knew she was on dangerous footing.

Nevertheless, her fear and concern for her sister overcame her reserve. "They say he was one of the men that shot Joseph and Hyrum." Charlotte was still, her mind racing.

"Oh Charlotte, don't go back to him. Please. I'm afraid for you. Come home with me. Papa is very sorry about what happened, and he and Mama pray for you every night. They want you back. Papa told me to tell you that he still loves you. Come back with me, Char." Annie grabbed her hands and clung on tightly. "You don't know Jack, really. I'm afraid of him, and I have been ever since I first met him. There is something dark and scary about him."

Charlotte's impatience broke in. "Oh, fiddle dee dee! You've been listening to Papa too much, pet. Jack didn't shoot Mr. Smith and his brother. Why he was way up north that day, arranging a horse trade. I happen to know. People love to make up stories about Jack because he's not a Mormon, and he's rich!"

The two girls sat down on the bank of the stream. The grass was warm beneath them, and a sweet musky smell came from the earth and flowers scattered helter-skelter through the grass. Charlotte put her arm around Annie and squeezed her. "Don't you think I know my own husband? I've lived with him for almost a month now, and I ought to know him. Jack's not a Mormon, and he's not religious, but he's not all bad."

They sat, each wrapped up in her own thoughts, tossing flower blossoms into the steam. Finally Annie said, her eyes brimming, "Then aren't you ever coming home, Char?"

"No, not to Papa's home. I have my own home now."

Annie burst into tears and threw her arms around Charlotte. "But I love you, and I'm so awfully lonesome without you, and nothing is fun anymore. Everyone is so quiet and sad, and Papa never laughs like he used to. Mama doesn't talk, hardly at all. I hate it there without you. I don't want you to go away and never, never come back. I always leave room for you in bed next to me, just in case you decide to come back some night. And I lie awake and cry. I miss you, Char," she wailed. "Please, please come home."

Charlotte was crying with her. "I can't, I just can't. I said I'd never see Papa again, and I won't go back on that. All those years of yelling and whipping were too much. I'm a woman now, and I won't stand for that anymore. I almost hate Papa when I remember how he humiliated me in front of Jack and the whole family. I would come home, if I could, for your sake, but I can't. Even if Papa weren't there, I belong with Jack now. I'm married to him, and I can't just go off and leave him."

They were quiet for a long time holding hands in their laps, lost in thoughts of happier days. Finally Charlotte's thoughts turned to more recent days.

"Why do people say that Jack shot the prophet?"

"Brother Taylor was there. He was in the jail with them. He was shot, too, but the bullet hit his watch and didn't hurt him much. He said he saw a good many of the mob. They were mostly from the neighboring town, and Jack was one he recognized. He said he heard him yell, 'I'll see you in hell, Smith' just as he shot. And later, when the mob left them for dead, he said he was crawling to the window when he heard hoofbeats and looked out. He said Jack was riding away on a palomino that he's been seen on before."

"He was sure it was Jack? How does he know him?"

"He's seen him in town and seen him with you. Yes, he says it was Jack for sure."

"Well, I don't believe it. He bought some horses up north that day."

Annie didn't answer. Shortly, Charlotte said, "It's getting late, pet, and I have to go home. Write me and tell me when you can come and visit me. You could at least come and stay all day couldn't you?"

"I'll write you. Maybe in a couple of weeks I'll have Papa talked into it."

But it wasn't a couple of weeks, it was four months before Charlotte heard from Annie again.

CHAPTER 4

Charlotte had written several letters home those first two months, and sent them by Jimmy John, asking him each time to wait for a reply if her sister wanted to answer it immediately. Each time the boy came back and said that Annie had read the letter and told him not to wait, that she had work to do and would write later. Finally Charlotte's feelings were hurt and she ceased writing. She told Jack about it, and he said he supposed that Patrick wouldn't let Annie answer her mail.

There were signs that the Mormons across the river were making plans to leave. They had been threatened many times with extermination from angry mobs and arsonists. So they were getting ready to leave their beautiful city, Nauvoo. They were trying to sell farms, lands, animals, and put together the provisions necessary for a long trek. Word had it they were headed west, clear across the Missouri River, maybe even out of the United States. Charlotte wondered if her parents were going. She supposed they would. They always had. Still, neither her mother nor Annie had written to tell her good-bye and she knew, despite all that had passed between them, that they wouldn't leave without seeing her one last time.

She and Jack had been busy, indeed. Jack had put the hired men to work helping him and Charlotte build their new house. Within a month they would be ready to leave the little cabin and set up housekeeping in their new home, palatial by any standards that Charlotte had ever had. It was a two story imitation of the gracious Southern mansions she had glimpsed in passing through Missouri. She and Jack had been in immediate accord as to the style of their new home. He had traveled much more extensively than she, from Chicago to New Orleans, and he, too, favored the open, spacious elegance of the Southern mansions. Right now they had only the bare shell of the home up, but eventually the bedrooms would all fan out from the top of the curved staircase, joining the main floor with the second. Upon entering the huge double doors, one emerged into a light and airy entry, large enough to accommodate several guests at once, and found the eye drawn up and up the wide, curving staircase to the right. The wood railings Jack had ordered were of rich mahogany, and, eventually, the steps were to be covered in Chicago's best tapestry. Jack particularly liked the staircase. It gave him an advantage over the guests that he would entertain. He knew instinctively that he would gain immediate control of the moment as he descended and visitors were forced to look up in anticipation of his arrival.

Charlotte liked the windows. She had insisted on a light, bright home. They had designed it together, with row after row of windows. The parlor, to the immediate left of the entry, was to be dominated by floor to ceiling drapery, falling in graceful folds of pale green and drawing attention to the dozens of panes that formed the wide, wide windows. Here Charlotte would, one day, sit with her babies, rocking them, tirelessly entranced by the patterns of glass and the small halos it cast upon the walls and floor.

Adjoining the parlor, but secluded by virtue of the high, carved doors separating the rooms, was the dining hall, large enough to seat thirty guests. Charlotte had argued with Jack about that. Thirty guests indeed! There weren't thirty genteel people in Montrose to invite. But Jack had a greater vision and he insisted, even though the room would be ridiculously large for the two of them at first. So, for years to come, they ate at one end of the great expanse of oak table, Jack at the end, and Charlotte at his left so she could watch the fire in the stone fireplace across the room.

If the parlor was Charlotte's room, this, the dining hall, was Jack's. He had chosen oak flooring, heavy oak dining table, a long oak mantel that ran the width of the fireplace. Huge rocks mounted from the hearth up the wall to the ceiling, and a wooden rack affixed to the stones, boasted his best rifles. Tucked into the corners of the room were Jack-sized chairs, overstuffed with padding for comfort, upholstered in dark, wine red tapestry matching the drapery for the window at the end of the room. Nothing in the room was small. It was the room in which he would bring his friends to relax, to fill up on the good whiskey he loved, and to get uproariously drunk when he felt like it.

In the back of the house, were the kitchen and the maid's room. Here, there was not only a large open hearth for roasting meat and stewing large kettles of soup, but also a wood-burning stove which insured that the room would be toasty warm in the coldest winters and sweltering in the summer.

Upstairs would be five bedrooms. Charlotte and Jack's room was directly over the downstairs parlor, sharing with it the chimney, and boasting a fireplace of its own. Two large mahogany wardrobes contained their clothing. These Jack brought back from St. Louis, where he acquired them from a ship's captain. The doors were elaborately carved, and the style was gracefully French. It was the only delicate thing Jack liked, and he told Charlotte often as he ran his hand over the smooth, glossy wood, that it reminded him of her. They slept in a large four-poster bed covered with the quilt her mother had sent in the trousseau, and Charlotte slept on the side nearest the windows, so that the early morning sun fell on her first, starting her day with sunshine. At first the walls and floor were bare, but, eventually, Charlotte had the walls covered with a silken fabric, finely patterned in yellow and white flowers, and the curtains that billowed at the windows were pale yellow. It was a hopeful room, a room where they should have loved much and forgotten their differences, for that was how Charlotte designed it. But her hopes never were fulfilled.

In October Charlotte was three months pregnant and began growing anxious to share the good news with Annie and Margaret. How Charlotte would have liked to have Margaret near now to confide in and to answer questions. Her body was becoming a stranger to her. Even her emotions and temperament were changing, becoming gentler, less contentious. She found herself asking Jack, night after night, to come sit beside her on the porch and look at the moon. Occasionally she even remembered snatches of the pretty poetry that Orin used to recite to her in his red-face, stumbling way.

She had watched Jack for weeks after her talk with Annie, trying to discover anything that would cause her to believe what her family said. She found nothing. Indeed, his temper was milder now than ever. He was in the best of humor, patient with her and with the animals, and Charlotte had hopes that the usual standoff between them was dissipating. He promised her a horse, any one that she wanted. She had only to pick one out and he would break it and train it for her. About a week after her meeting with Annie, five horses came in from the deal he had made up north. Amongst them, was a two-year-old chestnut colt. He was gorgeous, with long slim legs, graceful body and a fine, sensitive head, tossing proudly as he galloped round and round the ring. His coat was a rich amber, much like Charlotte's own hair, and his mane and tail a deep brown with reddish glints.

"That's him, Jack. That's the one I want."

"Charly, you must be crazy. That horse is worth better'n a hundred dollars when I get him broke."

"No, I'm not crazy. I want him. You promised me any horse I wanted."

"Yeah, but that's a valuable horse."

"So what do you want me to have, a nag?"

"Honey, you can have any horse here except that one."

"That's the only one I want."

"Charly, you're worse'n a plague. You can't have it."

"Ye said I should have a horse, Jack Boughtman, any horse I picked. Ye did na' say it had to be a worthless one. How would it look, yere own wife riding a nag?"

"I never said you had to ride a nag. Just not this horse."

She stamped her foot angrily. "If I can na' have this one, I'll ride a jackass and let ye see how the townsfolk laugh."

"I believe you would."

"Ye may be sure o' it." She was set. Her eyes were green fire.

"Well, then, if it means that much to you, I'll give you a two hundred dollar present."

"Ah, the price is going up. I thought you said he was worth a hundred."

"Two, my lass. Two hundred, if I know horseflesh. Are you pleased? I don't know another man that would give a woman a two hundred dollar present, especially a hard-headed, sharp-tongued spitfire like my woman." He kissed her while Emery looked on, embarrassed a little, but amazed that Charlotte stood up to Boughtman.

"All right, he's yours. What do you name him?"

Charlotte grinned and kissed him again, loudly, with relish for having won this round. Then quickly, she whirled around and dashed over to the corral where the horses were strutting about. She climbed up on the first rung of the fence and watched the chestnut pacing the ring impatiently. She could hardly look at him enough. He was perfectly beautiful, regal, a king, and he was hers. She sucked in her breath at the delicious exhilaration. "I'll call him Shannon," she said. "That's a good Irish name for a chestnut colt. And I'll break him," she finished, turning back to Jack.

"Now, Charly, you can't break that colt. He's as wild as I've ever seen. It'll take a man to do it. You have to show a horse who is master. A woman could never do it."

"I can."

"Nope. You're too soft on the animals. You have to be mean sometimes. I'll break that horse, and he'll be so gentle a year old baby could ride him."

"Did you give him to me?"

"Well, I said so, didn't I?"

"Then he's mine, and I want to break him."

Jack got angry then and kicked the fence post. "Damn it, woman, can't you get anything through your head? That horse would tear you apart if you even got near him. You make me mad sometimes, but I don't want a dead bride. They ain't so much fun. I'll break and train that horse or you don't so much as touch him."

"No, Jack," she said quietly, still looking at Shannon. "If he's mine, then he's all mine. Let me do it my way."

In the end, she did it her way. He was her horse all right. Shannon would do anything for her. Everyday she went out to the corral and stood up on the fence, holding an apple in her hand, calling to him softly, almost crooning his name. On the fourth day, he sidled over to her, looking carefully in the opposite direction to let her know he wasn't really paying any attention to her. But he took the apple from her hand. A week after that, she was able to stroke his nose, and soon she was standing inside the corral, calling him. He would come galloping into the corral, thunder around the ring several times, passing within a foot of her small, slight frame. When she didn't budge, he would stop short in front of her, pawing the ground and snorting. Charlotte stood her ground and persisted in extending her hand filled with goodies for him. The day finally came when she called his name, and the big horse came prancing to her outstretched hand, and suffered her to throw her arms about his neck and lay her cheek on his warm, silky coat. They were both in heaven.

Jack teased her about it continually. He watched her progress, slow by his standards, but as Shannon began to respond, then show obvious affection and gentleness toward the girl, Boughtman began to admit that, perhaps, her method might work. Even still he cautioned her, "You've yet to show him who is boss."

Charlotte only answered, "No one is. At least I'm not. I adore him and if anyone holds the upper hand, he does because I love him."

After the second week, there was no upper hand. The horse seemed to adore Charlotte in return and never approached her except at a slow walk, with his nose down, waiting to be rubbed. When she halter broke him, she used a special, soft, cotton rope that no one ever touched except herself. She would tie it with a double hitch under the chin, and only the slightest pressure told Shannon when he was to come and when he was to stop. She never had to jerk it tight. He was keenly sensitive to her every command. Finally she mounted him bareback. She spread her old cloak over his back, and Emery gave her a leg up, amid much protesting. Shannon was nervous and skittish at first, dancing around the ring, but he never reared or tried to buck her off. After much snorting and tossing of head, Charlotte's voice coming close by his ear settled him down a bit. She only stayed on a few minutes, then slid off. He went trotting free and eyed her from the other side of the corral. After a few minutes of calling and tempting him with an apple, he sidled up to her and gobbled down the apple, while she stroked his neck and nose, whispering her love words to him. Within a few days, she could mount him at will, and he seemed to enjoy carrying her about with her hands entwined in his mane.

The saddle and bridle were another hurdle to cross. They took another couple of weeks of coaxing and crooning and apples. In fact, though Shannon suffered the saddle patiently, he always responded with high spirits when Charlotte dispensed with it and climbed on his bare back, leaned down with her cheek against his neck, her hand in his mane, and urged him out into the fields for a lark. Before two months were out, she was riding him anywhere she wanted to. She eventually taught him to jump, to prance, and even to "shake hands".

Jack had been right. A man came down from Chicago to look over the horses on the ranch. When Charlotte came riding in on Shannon after their daily exercise, he offered Boughtman one hundred eighty five dollars for the chestnut horse. Jack told him it wasn't for sale. It was a present to his wife. The man looked at him like he was crazy and offered him an even two hundred dollars. Jack looked up into Charlotte's eyes, as she sat astride Shannon, and turned it down.

Early in November, Annie came. It was about 9:30 at night and Charlotte was home alone. Jack was gone, as usual, on business, he said. She was painstakingly crocheting in front of the fire, listening to the wind howl outside. She had known for three months that she was pregnant, but hadn't found the right time to tell Jack. She had begun crocheting in order to become proficient enough to make the baby clothes. She hated needle work, as always, but she found some enjoyment from the memories it brought of the many evenings the O'Neill women sat by the fireplace sewing, mending, and making lace for their dresses, and tablecloths.

She heard hoofbeats and thought it was Jack, hurrying home before the storm broke. Having resolved to have a talk with him and tell him about the baby, she threw open the door to find Annie instead.

"Annie! Is it really you? Come in, come in, get out of the wind. What are you doing here in the middle of the night? You look blown to pieces."

Annie came cautiously into the large living room, peering about anxiously. "Are you alone, Char?"

"Yes, why?" She saw in a moment that something was very wrong. "What is it, pet? Are you in trouble?"

Annie turned to her and gripped her shoulders. "No, but Papa is. They've burned us out and tarred and feathered Papa. He's bad, Char. They got it in his eyes and blinded him, I think. Mama and Papa are over at the Logan place. They're trying to get the tar off with turpentine, and he's hollering terrible. I heard Mr. Logan say he doesn't think Pa will live through it. They beat him first and dragged him behind a horse before they poured the tar on. He's most out of his head, and he keeps calling for you and crying. I never saw him cry before. Please, won't you come and talk to him. I know he'll die if you don't." She burst into tears.

Charlotte was overwhelmed. She had grabbed Annie now with both hands. "The farm, is it all burned out?"

Annie sobbed a yes. "Everything is gone."

"What'll you all do? Who did it?"

"I guess we'll go west with the others. They are getting ready to leave as soon as they can. We thought we'd have the money to go soon. Papa had a buyer for the farm and one for the store. But now I don't know if he'll be able to sell it."

"Annie, why didn't you tell me all this before?"

"I did. I wrote you every week, and when you didn't write back, I thought maybe you were mad about what I said about Jack last summer."

Charlotte stared at her. "Well, I never got your letters. Who did you send them by?"

"Papa sent them for me by that Jimmy John boy that works here."

"He never gave them to me. Did you get mine?"

"We never heard a word from you."

"Let's go, Annie. Let me get my cloak and put on my boots." Charlotte was angry now. Fury began to boil as she realized, without a doubt, it had been a conspiracy to break all her remaining ties with her family. Jimmy John would not have withheld her mail except on Jack's orders.

It didn't take them long to get to the ferry. With the storm whipping up white caps on the river, the man was reluctant to take them across; but he did it anyway when Charlotte told him who her husband was, threatening him with Jack's well-known displeasure if he refused. Once across the river, they rode with the wind at their backs and, before long, were at the Logan place.

Charlotte jumped down off Shannon and told Annie to walk out the horses; then she burst into the farmhouse. Only the light from the fireplace and one dim lantern lighted the drab room. Patrick was laid upon Tom Logan's bed, stripped almost naked while her mother bent over him, patiently swabbing him with turpentine. He was exhausted and weak, and hardly made a sound when the stinging liquid touched his wounds. Margaret straightened and turned to her as Charlotte stood gaping from the doorway.

"So, you came."

"Mama, is he going to die?"

"I don't know. Does it matter to you now?"

Tears began a slow concourse down Charlotte's cheeks. "Yes," she whispered. "It matters." She went to the bedside and looked down at the red, Irish head, black now with tar, his eyelids stuck together, the remains of chicken feathers scattered about on him like down. He was a nightmare, a spectre of the night. "Papa, it's Charlotte. I came when I heard. Papa, can you hear me?"

He didn't answer. Margaret spoke. "He can't hear you. He's too far down." Bitterly she asked, "Does he look like your father to you?"

Charlotte shook her head, no, tears dripping off her cheeks onto Patrick's. Margaret sighed and turned back to her job of wiping. "Then I guess you'll get your wish, of never seeing his face again. Even if he lives, I doubt his face will be recognizable, for his nose is broke and most of the teeth gone from his head, too. He'll be blind, like as not."

"I'm sorry, I'm sorry, Mama," Charlotte was weeping alone, her mother too intent on Patrick's pain to care about her daughter's. "I'm sorry, Papa. Papa, can you hear me? Please, please hear me. I came back. I came back. It's me. It's Charlotte. Can you hear me? Oh, please hear me, Papa."

Pat moved his arm, and winced. Then he struggled to open his eyes but the eyelashes were still caked in tar. Finally, he whispered. "Charlotte? Where's me Charlotte girl?"

"Here, Papa, right here beside you."

He drew a deep, shuddering breath and a tear squeezed from between black eyelids, slipping over the tarred cheeks. "Charlotte, ye must not go back, lass. Ye must stay with us. Promise me ye'll not go back."

"Why? Why, Papa? You know I'm married these six months. I can't leave Jack."

"Ye must not go back to him, lass. He'll harm you. Promise ye'll not go back."

Charlotte looked up at her Mother. "I don't know what to say, Mama. I can't promise that, but I don't want to upset him." Her mother was like stone. Charlotte would get no help from her. She turned back to Patrick. "Why, Papa? Why can't you accept Jack as my husband?" He struggled to raise his head, but the effort was too much, and he fell back on the pillow. She had to lean down, her ear next to his lips, before she could make out his words.

"It was Jack what did it."

She jerked back as though she had been burned. "What do you mean, it was Jack?"

He nodded ever so slightly. She looked back to her mother, stunned. Margaret moved her head wearily. "He's talked of nothing else. Jack led the mob. There were a dozen men. I saw them come before your Papa shooed us out of the house and off across the fields to the Logans. He was on a palomino. Patrick wouldn't shoot him because he was your husband. He should have." Her voice grew louder and bitter as an Illinois winter. "He should of shot the yeller coward. Lord, how I wish he'd shot him and sent him straight to Hell, where he belongs with the rest of Satan's band."

Charlotte had risen from beside Patrick's bed. "I don't believe it."

For the first time in her life, her mild, cool mother shouted at her. "You never believe anything, do you. You think you're the only one who knows anything at all. Do you think I'm lying, too, and Annie, too? She saw him, you know. Do you think your Papa on his death bed is lying to you, you foolish, foolish girl? The man is evil. He is evil, I tell you, and you'll be lucky to escape him with your life."

Margaret was a slow-burning ember but the heat was white hot now. Charlotte didn't know what to say. She was more afraid of Margaret's anger than Pat's, and more afraid of what she had just heard than she dared admit.

"Annie, did you see him?" Her sister nodded, yes, slowly.

Charlotte made her way, painstakingly, to the doorway of the farmhouse. Everything was falling apart around her, and she was frightened, trying to make some sense of it. Could she believe that Jack would do all those terrible things? Could he do this to her own father? She was soon to have his child. She had to believe in him. She had to.

Outside she found Shannon, big and comforting as always. She put her arms around his damp neck. He snorted and whinnied to her, knowing something was wrong. Tears dropped heedlessly onto his lathered neck. Annie came up behind her. "You aren't going back are you, Charlotte?"

"I've got to," she said heavily. "I'm pregnant, Annie. I'll have his baby in May. Besides, I'm still not sure I can believe all of that. Maybe there was a mistake. Maybe he was there at first but left before all that happened to Papa. I have to ask him; I have to give him a chance to explain."

Just then their mother came out. "He's asleep now. I thought we'd better let him rest a bit. Lord knows if we'll ever get all that tar off and his eyes open." She came to Charlotte and stood beside her, looking intently into her face. Margaret knew Patrick loved this girl best. At times in her life, she had been almost jealous of Charlotte and the wealth of love Patrick had for her. Now she only pitied her.

"It means a lot to him to know you're here. It broke his heart the way you parted. He hasn't been the same since. Won't you come and go with us out west? We're all afraid for you."

Charlotte threw her arms around her mother and sobbed. "I can't, Mama. I'm pregnant, three months gone. Besides, Jack would never let me go, even if I tried. I'm scared, Mama. He's never been like that with me. He's good to me. We've been happy. I never knew what he was doing all the times when he went out at night. I don't know how to face him now, but I have to at least see him once more and give him a chance to explain."

The three of them stood talking and comforting each other for a while. At last Charlotte said she had to leave. She went back inside one more time. Patrick was resting, only semi-conscious. She stood beside his pitiful frame, covered lightly with a thin blanket. Kneeling beside the cot, she reached out one soft, cool hand and lightly stroked his blackened cheek. He moaned and stirred slightly. She whispered softly, "Oh Papa, I do love you, so, so much." Tenderly her lips touched his brow, then his eyelids, then his cheek. The sickening taste of tar turned her stomach. His hand moved weakly on the covers, and she covered it gently with her own. Bowing her head over his chest, she whispered a prayer, "Dear God, I'm sorry for everything, for the way I am, for all the shouting and anger and the hurt. Just let him know I love him. Please, please tell him I love him."

After a few minutes she put her head lightly on his chest listening to the struggling heartbeat. Tears pooled in her eyes and wet the coverlet beneath her cheek. "Patrick O'Neill, don't die now. Not before I can tell you how sorry I am. I broke your heart and you broke mine, but oh, didn't we love each other." But he was sleeping too deeply to even stir now, and finally she rose, full with sorrow, heavy with a great river of tears struggling to break the dam. Reluctantly she made her way back to the porch and her mother.

As she mounted Shannon, she told Margaret and Annie she would be back the following day to see Patrick. Her mother pulled her head down and kissed her, whispering that she loved her.

It was the last time Charlotte saw any of them.

Shannon covered the distance easily in spite of the workout he'd had an hour before. The big red horse pounded the, hard, icy dirt with hammer hooves, his legs stretching out smoothly to eat away the miles. Charlotte was no more trouble to him than a cloth doll strapped to his back. The country roads were easily traveled, and she was not afraid of the dark or anyone who might be abroad. Shannon could easily outdistance any rider who might give chase. But her thoughts were not on her own danger. It never crossed her mind that anyone would try to harm her. She was completely consumed with visions of Patrick, her big, bluff father—a man she had loved and fought with, a man who had her greatest respect and her complete disrespect, a man she had wanted never to see again. Now she had seen him, or what was left of him, and all the whippings she had endured at his hand had not hurt like seeing him broken. She thought of Jack. He had broken her father just as he often did horses. The more stubborn an animal, the more extreme measures he took to break it. So he had done with Patrick.

The wind rushed at her. It was bitter and promised winter's snow soon. It whipped her hair and cloak, lashing her own anger and pushing her toward the river and home as fast as Shannon's powerful legs could carry her. Soon they reached the river, and although the ferryman was still there, he stoutly refused to tempt his luck again. The river was getting angrier by the minute. The wind had whipped up waves and topped them with white foam. Clearly, any vessel on the water would be in peril. Nevertheless, Charlotte's will was impressive. So was her money. She offered him two of the four gold pieces she had brought with her. The man looked into her determined face. He watched the big chestnut horse, studying the probability of the animal getting skittish and sinking them midway.

"I can't do it, ma'am. That there horse'd sink us afore we was half across."

"No he won't. Shannon will be perfectly easy with me standing at his head."

He shook his head, streaming wet with the torrents of rain. "I dunno. I can't chance it."

"He was all right before when you took us over. He didn't move a muscle."

"Yeah, but the river wasn't this bad then. It's crazy to try to cross now. Look at that water, ma'am. If you are Boughtman's wife, he'd kill me hisself if anything happened to you."

"I have to get home to my husband, and he'll skin you for sure if you hold me here. I have one more gold piece. It's more than you make all year. Let's go now while we can still make it." She had pulled Shannon aboard the small barge, and stood stroking his neck and nose.

The ferryman still stood, uncertain but greedy, until she said, "I'm letting off the lines. Obviously she meant it, and he quickly snatched the gold piece from her hand, grabbed his pole, shouting above the storm, "God help us." At both the front and back of the barge were heavy ropes which were attached at the other end to a huge pulley, which a partner worked on the opposite bank, helping the ferry across the river along a stable path.

The river man had been right, the crossing was going to tax everyone's endurance and determination. The barge was pitching to and fro, up and down like a cork in the churning waters, it's pulley lines taut and straining as they never had strained before. Waves rolled easily across the deck of the vessel, and Charlotte, by this time, was thoroughly drenched. The ferryman was cursing, working his pole, and appealing to the Almighty, all in the same breath. Charlotte stood beside Shannon, talking softly to him, stroking him, and the horse seemed much calmer than the man who had worried about his temperament. At last, amid much pitching and tossing, the opposite shore was approaching.

Charlotte's mind raced ahead to what she would find at home. Behind all the anger and accusations that had consumed her was a tiny hope that Jack would be in bed, calm, waiting, unaware of any of the trouble that night. It was possible that Papa was simply delirious. Annie could have been mistaken, like a little child who has been told there is a boogey man, and so has imagined a real one.

The miles to the ranch vanished under the steady thunder of Shannon's hooves. They turned in at the gateposts and Charlotte dismounted. The cabin was dark, but there were voices and a lantern in the barn. The dark and the boisterous voices covered her low, quiet approach. Shannon was blowing softly from time to time, but no one heard, and no one noticed her until she was framed in the doorway, the dark amber horse tired and sweating by her side.

She stood still, silently taking it all in. Jack and five other men, many of whom he had business dealings with, were sloshing turpentine around and wiping down hands, forearms, and boots. The cloths were black with tar, and the stench was everywhere.

Jack had his back to her and was congratulating Harold Wooding, a shrewd mouse of a banker. "I reckon you won't have to pay a cent for that land. It's worthless now."

"Yes, but you just wait three years, and we'll see what price that land'll bring when the last of these damned Mormons are rooted out." Wooding turned to see what was causing the hush, and when he saw Charlotte in the doorway, he ducked his head and moved to the other side of Jack. Jack glanced back over his shoulder, wondering what was going on.

She had never bothered to pin up her hair that night when Annie had first come and called her away. The ride back had tangled it and the rain had darkened it to a deep, reddish-brown. Her face was dead white, standing out like another dimension against her hair. Her cloak was wet and flung about her shoulders like an Indian squaw, and on her cheeks were raindrops, magnifying her freckles. Jack's heart leaped inside him for a moment. The most beautiful possession he owned. Then he saw her eyes, green fire, and her lips, thin straight lines in her face. He grew wary.

"Charly, thought you'd be in bed by now. What're you doing up?"

She didn't answer. The men began to shift around, tucking cloths into corners, eyeing each other, waiting to see what was up.

"Where ya been?" he asked. "You're all wet."

Still she didn't answer, but her fingers moved slightly to her right side, and she touched the crop that Jack often used on his horses. In the darkness and the shadows, Jack missed the movement.

"Is something wrong? You all right, Charlotte?"

Her voice was slow and soft. "Where have you been, Jack, and what's the turpentine for?"

They all looked at each other covertly. Jack was getting nervous.

"Why, uh, Rex there had some tarring to do on his new house."

"Until midnight? In the rain?"

"Yes, until midnight." His voice was rising as he grew embarrassed at being questioned in front of his cronies. "You'd better git in the house, you'll be catching the fever."

"What's in the sack . . . feathers?" she asked, not letting up.

"I said, git in the house, damn it. We'll talk about it later."

Her voice broke the uneasy stillness with a shout of outrage. "We'll talk about it now! We'll talk about me father now! We'll talk about the farm and me family and me letters, right now! We'll talk about poor Patrick O'Neill, blind and dying because of ye!" She took two lightening steps toward him and brought down the riding whip on his shoulders. The men were stunned; Jack was enraged. He grabbed at the whip, wresting it from her.

"All right, we'll talk. Yes, they're feathers, and this is tar, and what else do you want to know—how long it took the farm to burn down?"

"Ye are a fiend, Jack Boughtman!" She screamed at him, eyes narrowed, burning him with the hate in her heart. "Ye and all yere friends, disciples of the devil. Me mother was right. He should have shot ye and killed ye dead. 'Twould have been a service to the world. 'Twould have been a service to me." The tears so long pent up inside were breaking now, but never a trace showed on her cheeks. They came streaming out in a bitter torrent of words. "He could have killed ye! He saw ye coming. He could have saved everything, and shot yere head off, but he didn't because ye are my husband." She spit when she said the word 'husband'. "Before all heaven, I wish he had shot ye and all yere devils with ye."

His hand shot out and brought blood to her mouth. He was livid, deadly cold. "Shut up, woman! You've raved enough. He's a Mormon, even if he is your father, and he deserves no better than any of the others. We've been doing what we have to to drive them out. They're a pestilence, and we aim to be rid of them. Now, you git in the house, and don't show your face again, except to apologize to these men."

She could hardly believe it. She stared at him in awe of his unspeakable nerve. "Ye must be sick in yere mind. I should apologize to these devils?. After they have burned out me family, most killed me father, and laughed about it? Oh, yes, I heard ye laughing when I come up. Heaven help me if I ever so much as speak a civil word to them, much less apologize." She wrenched away from his grip. "And ye," she whispered, "ye have deceived me ever since we were married, probably before. Ye have kept letters from me and made me to think it were Papa's fault. Ye have played the loving husband to me, and all the time lying, killing, burning behind me back, and me defending ye to me family."

The men had begun moving toward the doorway, but Shannon was still there, blocking it, waiting for his mistress. She saw him, and a resolution came to her. "I'll not go back to yere house. I'll not live with a man who is capable of such cruelty and lies. I go back to me own, and ye can tar me with them the next time." She grabbed the reins of her horse, springing quickly up on his back, and started backing him out of the barn. But Boughtman was on them before she could clear the barn door.

"The hell you will! Git down from there. You're not going anywhere. Like it or not, you're my wife and you'll do what I tell you."

She went wild then, and started flaying him with the crop. The horse danced nervously underneath her. Jack dragged her off Shannon, and she screamed at him. She turned into a wild thing—a mountain cat, scratching, biting, gouging at his face—and after a few minutes, unable to contain her, he flung her away, slapping her hard with the flat of his hand. She fell a few feet away, spitting blood on his boot. "I spit on ye, I spit on ye, Jack Boughtman, Satan's own son!"

He lifted his boot in blind fury, lost his balance and brought it down just below her ribs in a sharp glancing blow. Williams and Stacy, two of the men, moved alongside Jack, cautioning him not to hurt her. "She'll calm down," Stacy said. Just at that moment Emery came running up in his nightshirt, having heard the shouting. Charlotte was doubled over on the ground, and Shannon was nosing her as a mare might her foal.

"My Lord," he said under his breath. "Mrs. Boughtman, Mrs. Boughtman. Are you all right?"

"My stomach," she groaned. "It's the baby. I think it's the baby!" Involuntarily she let out a gut cry of pain.

Jack stood, staring down at her. "What baby?" he asked dumbly.

Emery looked up at him, reproach on his face. "Yours. She's been pregnant for three months." He picked her up carefully and carried her, like a chinadoll, toward the cabin. Then Jack came to life. He kicked open the cabin door and led the way to the tiny bedroom.

"Lay her here. Is she going to be all right? Will she lose the baby?"

"I don't know. I'm going to get my wife."

Charlotte grabbed his arm and held it tight. "Don't leave me here with him. Take me with you, Emery. I never want to come back here again. Please, don't leave me, please."

The ranch hand looked up at Jack, his face awash with emotion. Then, regretfully, he patted her hand and said, "I can't take you, Mrs. Boughtman. But I'll be back, and Lily with me. She'll take good care of you. Don't worry about anything, your baby will be all right. Just stay put now and don't move."

He left, and Charlotte tried to struggle off the bed. Jack sat down beside her, easily pinning her arms against the pillow. His voice was as gentle as a kitten. "You've got to stay here. I'm sorry, Charlotte, I didn't know."

The hatred in her eyes seared him, and disgust showed through the gray pain etched on her face. "It wouldn't have mattered if you had known. Nothing matters to you." She tried again to get up, and again he gently restrained her.

"Now you're gonna hurt yourself worse. Lay down and be still."

"I'll never be still with you touching me."

"If I go set in the rocker, will you promise you'll lay quiet?"

"I won't promise you anything."

"Then I stay."

She was quiet a moment. Finally, "I'll lie still. Just take your hands off me and let me alone."

He got up, standing silent and tall, looking down at her, wanting to say something but not knowing what to say. Then he pulled the rocker up into the doorway and sat down, looking at her, as she lay spent and drained on the narrow bed. For a long time, she lay with her eyes closed, and Jack thought she was asleep. He was sick with fear for her and with grief over what his temper had done to her. If he could be said to love anyone or anything on this earth, it was Charlotte, strong-willed, proud and stubborn, with the beauty of a wildflower. All the rest—the burning of her family home, even the tarring of Patrick—those were unimportant to him. They had no meaning, could not touch him inside. Only she could. Only she could make him wildly happy one moment and wildly angry the next. Right now, all anger was vanished. Even his pride, being humiliated in front of his friends, even that was forgotten as he tirelessly watched her face, pinched in pain, white against her tangled, wet mass of hair. He sat, helpless in his love for her and his regret at what had happened.

After a while, Charlotte's eyes opened slightly, and her voice broke his reverie. "I'll always hate you, Jack."

He sat looking down and didn't speak.

She spent most of her pregnancy in bed. He had her watched continually, until word came that her family had gone. On the day of the first snowfall, she determined to walk to Nauvoo and join her family, but she never got farther than the end of the lane before pains and bleeding stopped her. Jimmy John found her lying in the snow with a little pool of blood around her. After that, Jack wouldn't even allow her outside the house. He sent a letter to her mother stating that she was remaining with him. The O'Neills, including a sick and blinded Patrick, left Nauvoo with the second wagon train in March. Pat died on the Iowa Plains, with the spring floods mixing the mud ankle deep on the Mormon wagons.

CHAPTER 5

He was a tall man with reddish blond hair—more red than gold these past few years—but let the sun have it's way and the gold leaped out like spun threads. He was standing beneath the huge elm tree at the head of the lane, studying the house before him. So many times he had done the same thing in the last eight years, stood and looked and tried to tune himself to the particular spirit of each home he approached. John Patrick O'Neill had just finished eight years of service to the Lord as a missionary for the Church of Jesus Christ of Latter-Day Saints. After his first four years, the Church authorities had sent word to him that he could come home. He replied that he appreciated the offer but was quite content in the work, and if the Lord wouldn't mind, he'd rather continue to serve.

Throughout New England and Nova Scotia he was known as "The Mormon". John had left Nauvoo a stubble-bearded farm boy, with huge, work-grimed hands that had never known anything but horses, dirt and a plow. In these last few years, they had known fish nets, scales, fine china, silken linen, dog kennels and thousands of hand shakes. He had taught the Mormon Gospel to hundreds of people, and by hook or crook, coaxing or thunder, he had baptized almost two hundred people himself. Most of those new converts had already beaten him to the Great Salt Lake valley, where he was now headed. His mother and Annie had a small place north of the city, and he was anxious to put his hand into his own soil again.

He had been curious about Nauvoo. His curiosity was now satisfied. He had seen it—the half demolished temple, the city all but deserted, the outlying farms overgrown. The charred remains of many homes were being eaten up by the weeds and underbrush, their black accusing memories conveniently disappearing with the eternal tears of spring and the fruits of summer. Of the O'Neill farm, there still remained the chimney and most of the back wall. He had walked around, trying to imagine, from his mother's account, the burning and the tarring and feathering. After the O'Neill anger had worn off, he had sat down on the hearth and smoothed the cold bricks with his fingertips. John Pat gave way to the eighteen-year-old boy inside. Here he had yelled at his bull-headed Irish father and received a good cuffing for his impudence. When it was over, he looked his father straight in the eye and said, "Thank you, sir," and Patrick had thrown his beefy, freckled arm around his son's shoulders and hugged him.

Charlotte had been disgusted. That he would thank Patrick for beating him was preposterous to her. "You have your way," her brother told her, "and I have mine. Papa was right to cuff me. I was wrong to speak to him that way. If I had gotten away with that, I doubt I'd ever have really obeyed him again."

How many times he could not remember, he had searched the farm over to find Charlotte cooling her willow-switched legs in the stream, or sitting high in a tree nursing in her mind the blows dealt her pride, as well as her backside. "Charlotte," he had said on one such occasion, "if you were my daughter, I'd whip you, too. You're disobedient and rebellious. What'll it take to learn ya?" He was sitting below her on a branch where the limbs forked. Charlotte could see he was really saying "I love you," as he looked up at her with his starry-blue eyes.

She had thrown a handful of leaves at him. "Oh you would not, Johnny. You love me too much. You'd never bring yourself to whip me."

He shook his head, "Pa loves you, too. He just shows it in his own way."

"Well," she had replied, "if that's what a switching means, you can take them all for me, and say thank you when it's over. As for me, a switching will never teach me thanks or love or even obedience."

Emerging from his reverie, Johnny had gone in search of the Boughtman house. It wasn't hard to find. He simply crossed the river and asked the first person he saw. Jack Boughtman was a well known fixture in Montrose. The house was surrounded by flowers, yellow and blue and various shades of pink. It was a beautiful two story building, and the windows were real glass. The chimney brick took up one whole end of the house. It was definitely a rich man's home. John Pat was slowly eating the end of a blade of grass, watching and wondering. What was she like after all this time? Was her husband at home? What should he say to him, the man that had murdered his father? He was trying to sort it out in his mind before he approached the house.

Just as he was almost ready to make his move, he saw the door open and a boy about six years old come out. He was dark-haired, dressed in overalls, and was obviously impatient. Suddenly he whistled, and Johnny became aware of fast-approaching horses hooves becoming louder and louder. He spun around, and there were two fast horses with riders hurtling toward him. One was a small palomino horse, with a very small girl about four years old urging him on, and the other was a huge chestnut animal, stretched out like a fine silken thread. Charlotte was astride him. It took no more than a second to recognize her, though she had changed considerably from the neatly gowned, neatly braided girl he had grown up with. Her hair was loose and flying, and she was wearing a pair of man's pants, faded and old, and a simple cotton shirt. One sleeve was cut off above the elbow, and the skin it exposed had seen a lot of sun.

He stepped out from under the boughs of the elm and into the sunlight. The big chestnut stallion was reigned in. He stopped almost in mid-stride, and was drawn up, snorting and dancing, just a foot away from John Pat. The little girl kept going and sang out triumphantly, as she turned her horse around in front of the boy, "I won. I beat you, Mother. I won, I won."

But Charlotte didn't hear her song of triumph. Charlotte was sitting still as a statue. Brother and sister looked at each other, both of them seeing ghosts of years passed. John Pat put his hand on the horse's bridle and smiled up at her, his missionary hat on the ground, and the bright flecks of light in his blue eyes glistening with unshed tears. Charlotte reached out a timid hand in disbelief and touched his face tentatively, as though he might dissolve beneath her fingertips. But no, he was solid, he was real. He was her own Johnny, looking now, after eight years, more and more like their father. The tears came rushing down her own cheeks, washing dust away as they came.

"Oh, Johnny," she whispered. "Is it really you? I thought you had gone back to Utah with the family long ago. I never thought I'd see any of you ever again." She slid off the saddle and into his arms, and they cried over each other like two little children, long orphaned, now a family again.

"Mommy," Ruby demanded imperiously, "Who's that?"

Charlotte took the reins of the palomino, drawing her daughter into the circle. "Get down, honey. This is your uncle Johnny. He's Mama's brother, and he's come to visit us. Isn't that wonderful? Come meet him. You, too, Matt."

She drew her two children beneath her wing and presented them to Johnny. Her face was shining. "These are my children, Matthew and Ruby. Aren't they beautiful?" But he saw two small, suspicious, petulant faces accosting him. Still, there was the little Charlotte in Ruby's set jaw and prim mouth, and Matthew had inherited her fine, gray eyes.

"Yes, they are lovely children, Char, so much like you." He said "how-do-you-do" to both of them and shook their hands gravely.

"Run along now, you two," Charlotte commanded. "Ruby, would you ask Sofie to put some food on the table? Your uncle and I will have something to eat. Matt, I think Emery is training that filly today."

"I know, Mama. I was already going there," the boy answered. "I was just waiting for Ruby to get back and go with me."

"Fine, you can both run along then." The children walked up toward the house, glancing back over their shoulders a couple of times and talking in hushed tones. John Pat could see they were suspicious of him.

If he had had any reservations about how he would be received by Charlotte, they were all dispelled now. She drank in the sight of him as though she could not get enough, and held on to his arm with both her hands as they walked up to the big house.

"Well, Charlotte O'Neill—oops, Charlotte Boughtman—you look younger than you did when we were kids."

She laughed happily, "Thank you, Johnny. I hope you're right."

"Mama always made you dress like the other girls, but even then I knew you'd have loved to dress in my clothes, if you could. I remember once you got switched for trying on my pants and shirt and riding Esther to the meadow."

She glanced down at herself in surprise. "Oh, the pants. I guess I kind of forget out here. I go riding most every day, and my dresses simply faded away after a few years. I guess Jack would buy more if I asked him. Emery's wife made these for me—he's our foreman." She asked anxiously, as the thought came to her, "Do you mind my looking like a boy?"

He kissed the top of her hair, "No, I don't mind you at all. You look like a queen to me."

The Negro servant, Sophie, was busily setting the great dining room table with cold cuts and fruit and milk. "Will this be all right, Ma'am?" she asked.

"Yes, Sophie, that's just fine." She took his arm proudly. "Oh, and Sophie, this is my brother, Mr. John Patrick O'Neill. Isn't he just the handsomest thing you ever saw?"

The black girl inclined her head politely. "Yes 'um. Pleased to make your acquaintance, I am." Then she giggled a little, and said over her shoulder, as she pushed through the kitchen door, "He shore am the handsome one, all right."

They sat at one end of the long oak table and scarcely noticed what they were eating. "Johnny, Johnny, tell me all. Tell me everything that happened since you left. Eight years is a long time to be gone on a mission. Have you heard from Annie, from Mama? How is everyone? Did you hear about Papa? Haven't you grown? You look about three inches taller than when you left. In fact, you look almost exactly like Papa used to."

He grinned at her excitement and obvious delight. "And you look three times prettier. I want to hear about you, and how you came to have such a grand house and two children and that gorgeous animal I saw you riding. He is a champion, Char. Is he yours?"

She was obviously pleased. "He sure is. That's my Shannon. We got him just after we were married, and I broke and trained him all by myself. He is gorgeous, isn't he! He's the best friend I've got. Really my only friend," she said pensively. "He hears it all, all the fights and bitterness. He has been nearly run to death a couple of times, when I was so mad at Jack I couldn't stand still. But, wait, I don't want to talk about that. You first. You tell your story first and don't leave anything out. Maybe later we'll get to mine."

She was an avid listener, and John Pat found himself deeply enmeshed in telling of the simple people he had worked with, fishermen of Nova Scotia, their salty tales, their stubbornness, their childlike faith. He talked on and on while Sophie cleared the table. He told about being tossed overboard by one burly fisherman whom he tried to convert at sea. He had followed McGregor onto his boat and preached at him while he put out to sea. Charlotte laughed at the image of Johnny floundering in the water, bellowing to the man that, if he didn't repent, he would lose his boat and all he possessed. McGregor had leaned over the boat and laughed at him until, after a few minutes, he felt the wind crawling up his back and noticed a squall blowing in. Soon his boat was tossing about in the waves, and water was sweeping in over the sides. Meanwhile, Johnny was tossing about in the waves like a soggy cork. McGregor pulled him in with his fish net, and Johnny came over the side, water-logged and shivering.

"Well," demanded McGregor above the noise of the wind, "Do somethin, mon. If ye can put a curse on, ye had better be able to take it off. Call on your Mormon God."

Even Johnny had been skeptical. He glanced around him. The boat was fast swamping, the waves tossing buckets of water into the boat as quickly as McGregor could toss one bucketful out. Johnny bowed his head and prayed aloud for the storm to cease. Nothing happened. He raised his head and grabbed McGregor's arm. "Do you repent?" he yelled.

"Sure I do," replied the Scotsman. "I'll repent of anything if this storm will abate."

"Will you be baptized?" John Patrick hollered into the wind.

"Sure, and as many times as ye please."

Johnny bowed his head again and fervently called on the Lord to still the storm. When he raised his head many minutes later, the winds had blown over them, moving out to sea, whipping up white caps yards away, then miles away, and, as McGregor dipped out water from the bottom of the boat, no more came in. The Scotsman and his whole family, and one third of the village, were baptized the next Sunday.

He talked on and on, telling one incident after another that had earned him the name of "The Mormon" in the fishing villages of Nova Scotia. Charlotte sat with chin in hand, hanging on every expression. He told her about the Governor of Maine who raised and bred dogs. He had spent three months waiting for a new companion to be sent to his area, living in the Governor's bunkhouse, helping with the dogs. He had eventually been invited to a small dinner party with the Governor and his friends. It was his first time handling fine china and linen. He broke a cup. He had felt very much out of place throughout the rest of the torturous evening, until the Governor turned to him and asked him about this man Smith the country had been hearing about. For three hours then, until after midnight, Johnny had told the story of the American prophet, Joseph Smith. Before he left two weeks later, he had baptized six who were at the dinner party and received a warm, standing invitation from the Governor himself to come back.

At last Charlotte interrupted him with a question. "Are you so certain, then, that you were right in baptizing all those people? What did you tell them about the Church?"

"Of course I'm certain. I've spent eight years doing it. I've seen changes take place in the lives of people that have brought them happiness. Oh, at first, when I was just eighteen, I was afraid sometimes, unsure, but as the years went on and I saw the gospel take hold of those people and lift them up, there was no doubt. Do you really doubt it?"

"Yes, I do, Johnny. What good did the Mormon Church ever do our family? We were chased out of so many places, burned out of our house, our own father was tarred and feathered because of it. I wonder, if God is so pleased with the Church, why does he allow this kind of suffering?"

He looked at her keenly for a moment, then said as gently as he could, "Perhaps He was separating the wheat from the chaff. He has never promised that His people would have an easy way. You find the mettle of a person when you put him through the fire. Some melt down like lead. Some turn into pure gold."

She didn't say anything. He asked, after a minute of silence, "Are you separated out, Charlotte?"

"Yes."

"But why? You know you never even gave it a fair chance. I remember when we were all baptized. You refused simply because Papa tried to make you do it. You didn't have any real convictions about it. You were just 'playing the mule', as Papa used to say."

"Yes, but I lived with it, like it or not, for six years, and I never found a good reason to join in all that time."

"You might have if you had humbled yourself enough to ask God if it was right."

"I prayed with the rest o ye," she answered swiftly. "Twas every morning and every eve, and I knelt beside ye, John Patrick, most of the time, so don't ye play holy with me."

He grinned at her brogue and grabbed a handful of red hair to tug. "Don't ye play holy with me," he mimicked her. "I see yere eyes be a gittin' green, Mistress Boughtman. Don't ye be gittin' angry with me, little lassie. I've me Papa's good, thick, right hand."

They laughed together, and he chided her lovingly again. "Sure you knelt with us every morning and evening, but where were your thoughts? In the meadow, I bet."

She smiled at him and admitted they were.

"Char," he said earnestly now, "I never had a real personal feeling about God either, until I went on my mission. I just felt it was right in my heart. It takes prayer, constant prayer, and living right and wanting to do what He wants, before you really get a personal feeling about Him. Just try it, will you? All the questions you've ever had will be answered. All the help you will ever need will be right there. God wants to love you, to help you. Just open up the door."

She stood up and motioned at him with her hand. "Oh, Johnny, stop. In the first place, I don't want a God that you have to call Master. I've already had two men in my life that wanted to be my master, Papa and Jack. Well, I'm my own master, and that is the only way I'll have it. And in the second place, if there is a God up there, He has certainly never done anything for me, nor ever will I suppose. If you want to believe, that's fine for you, but I can't. Don't you think I haven't prayed for His help before. The night they tarred and feathered Papa was one. I prayed like anything for God to strike Jack down dead on the ground. Do you think I'd be here today if I could have gotten away then? But Jack kept me here, and the family left without me, and God didn't help me at all. If He'd wanted me to be a Mormon, that was the time He should have done something. But He didn't, and I stayed here because I had to! Jack held me here, and I had his baby, and I hated him like I've never hated anyone."

Johnny took her hands in his and pulled her down on his lap. "Charlotte, has it been that bad? Tell me, sis. You can tell me."

She began to cry, the tears slipping down her cheeks quietly. Then she buried her head on his broad shoulder and sobbed while he soothed her. Finally, when she got her voice back, she told him about her last eight years.

After she was confined to bed, Jack was very gentle with her, even bringing her presents (which she ignored) and things for the baby. When Matthew was born, Jack claimed him. The midwife took the baby and handed him to Jack as soon as the infant started to cry. Jack paced the small cabin floor, laughing and talking to his son. After that, he relinquished the boy to her only for feeding and cleaning up. The rest of the time, Matthew was his. She could see that he loved his son and was tremendously proud of him. He kept the boy by his side from the time he could toddle about. Indeed, Matt seemed to be his chief joy in life. He remained kind to Charlotte, and at last she began to soften. They went on picnics and rides together, the three of them. Eventually, his love for her and for the baby began to soften the iron of her hatred. After a few years, she decided she had no other choice but to make the best of her marriage and put that ugly, horrible night of the tarring out of her mind. She would never really forget it, and she could never love Jack the way she had wanted to love, but she could try to make her life happy. Within eleven months after Matthew, she was pregnant again, this time with Ruby. Jack was somewhat less exuberant about a little girl, and Charlotte had her heart full of a baby this time. Ruby was all hers—for the first two years anyway— and she found for the first time what it was to give and receive love openly and fully. She began to want that with a man.

She had, one evening, gone to Jack in her yellow wedding dress, put away for almost three years, and opened up her heart, hoping he would have learned some tenderness from his love for Matthew. She had spread layers of quilts upon the floor in front of the big hearth, and he had sat there with her, talking about the first time he had seen her in the meadow. In the soft, gold cocoon of firelight, she had hoped for a more tender love between them, but Jack took her as selfishly and roughly as always, though she begged him to be gentle. But he knew no other way and was too proud to try to learn. And so it was between them, always, until finally she no longer offered; he only took.

At first he had tried to get her to accompany him to the dinner parties of his friends. She refused steadfastly. She would not associate with the men that had persecuted her family. It was an insult to her, and she refused to lick their hands like a whipped dog. When she began to soften toward him, Jack tried again to get her to entertain his friends. Still she refused. Finally, he went against her will and called a big gala affair to be held at his brand new, showy home. He put Sophie, as well as two other colored women, to work cleaning and fixing up the house for the grand event. He purchased fine china and silver for the occasion. Charlotte refused to discuss it except to say to him, "Jack, if you do this, you'll be sorry." He brushed it off.

The night of the party, she would not come down. He greeted the guests alone, growing ever angrier with her. At last, he stood at the bottom of the stairs and called up to her, "Charlotte, your guests have arrived."

She appeared, then, at the top of the stairs, in the expensive, gold taffeta gown that Jack had bought her. In her right hand was a bucket of feathers. In her left hand was a bucket of warm tar, and smeared over the beautiful Chicago gown was a wide swash of fresh tar. She advanced down the stairs, calling out as she came, "Mr. Wheelwright, Mr. Williams, Mr. Stacy! I wanted you to feel right at home, so we are supplying your favorite playthings, tar and feathers. Mrs. Williams, so good to see you. Perhaps you didn't know how adept your husband is with tar and feathers. My father would be happy to tell you. Unfortunately, he died after he was tarred and feathered and run off his land. Mr. Williams, surely you would be happy to show us your skill with the tar."

Jack got to her before she could go further and struck her angrily across the mouth. The women gasped. Charlotte remained perfectly still, staring into his black eyes steadfastly.

"Get back upstairs," he hissed.

"I told you you'd be sorry," she said in full voice. "I will not entertain thieves, murderers and liars in my home. If their women don't yet know about their midnight exploits, then it is time they did."

He took her by the arm, his fingers digging into her flesh, and half dragged, half lifted her up the stairs. He pushed her through the door of their room and locked it. When he got back downstairs, his guests had begun to leave. They stood milling around in the driveway, waiting for their buggies. He knew the damage was considerable, so he moved amongst them apologizing for his wife. It was years before some of the women would allow him at their dinner table.

As Jack became a wealthy horse-trainer, dining with the Springfield socialites in his ruffled shirts and cravat, Charlotte became a recluse. Her dresses fell into rags, and Jack would buy her no more.

"I bought you a fine gown, and you cared only to ruin it and me too," he said bitterly.

"I lied," she told Johnny. "I have not simply drifted into this kind of dress because it is convenient. I dress this way because it's all I have. I go nowhere but the ranch or the meadow. I haven't been into town for four years, and no one from there comes here. The only friends I have are Emery, our foreman and his wife, Lily. She made me a dress a couple of years ago, and it disappeared from my closet the next day."

She sat on his lap, looking down at her hands, rough and dirty. "Oh, it's not that I care a whit for those simpering women in town or these vile people Jack calls his friends. I wouldn't care if I never saw them again, but I don't see anyone else either. You are the first visitor I've had since that 'party'. I would have been almost as grateful to see you if you'd been a stranger wandering onto the land. The children used to be some company, but they go mostly with Jack now. He takes Matthew everywhere he goes, even on business, and Ruby is gone almost as much. If she's not with them, she is off wading in the stream or exploring the ranch world."

John Pat had listened intently to the whole narration and heard the loneliness in her voice. "I'm going out to Utah, Char, as soon as I leave here. I want you to come with me. Mama and Annie have a little place just north of Salt Lake City. Part of it's mine, and I'm going to claim it. My part is yours too, until you find a good man you want to marry. If you don't marry you can stay on with me forever, as far as I'm concerned."

For a moment her eyes shone and her face was all alight. Then reality came back. "How can I go with you?" She shook her head. "You forget I'm already married. Besides, I can't leave my children, and Jack would kill us both if I tried to take them."

"All right," Johnny said. "Now it's my turn to get the Irish up. I've heard all I care to hear about Jack Boughtman. I've only met one man in my life that could lick me, and I'm anxious to try out any comers. If Boughtman wants to try his luck, I'm willing, but you can do anything you want to and that includes coming with me. When a husband fails in his duty to protect and care for a woman, then her men kinfolk take that responsibility over. Divorce is usually a terrible word, but for you there is no other choice. Come on, Charlotte. You won't even have to pack anything."

"I can't, Johnny. I'd never be happy in your Mormon city. It would be just the same as it was before. I'd be a misfit as always."

"Not if you were baptized."

"Oh, John Patrick, are you going to do what Papa did and try to hound me into it?"

"Have I ever?" her brother asked. "I won't hound you. But I would think that after all that has happened you would begin turning to God."

"Turning to God! I told you. I did. And He turned away!"

"I'm not talking about asking Him to strike someone dead. I'm talking about asking for guidance, for strength that you need."

"When God gets ready to give me a hand, I'll be ready to accept it, but so far He hasn't done a thing for me."

"He sent your spirit to earth and gave you life."

"Big favor! He sent me down here, gave me a push, and said, 'I'll see you in about sixty years.' I haven't seen or heard from Him since. Not everyone is as favored with manifestations as you seem to be."

"Not everyone is as hard-headed as you are either, my pet." He grinned at her.

"Well, be that as it may. I won't be baptized to something I'm not convinced of. I wouldn't fit into Mormon life, even if I somehow got away from Jack. There's really no place for me, Johnny."

"All right then, I won't go to Utah. We'll go back to New England, and you can keep house for me there. I like it there as well as anywhere."

She smiled at him tenderly. "John Patrick, you always were as big a fool as your father. I'm not going to let you give up all your plans for me."

"What, give up? I'll just make some new plans."

But Charlotte shook her head. "I would only be in the way eventually when you took a wife," she smiled wryly, "or two or three."

John sat looking at her thoughtfully. "That's what has really bothered you, isn't it?"

"I haven't thought of it for years," she answered. "But yes, it is. I could never share my husband. I thoroughly disagree with polygamy, and think it's inhuman to ask any woman to put up with it. I'll never know how they do."

"You couldn't accept it even for the security of a good home, a kind father, a gentle man who holds the priesthood."

"I don't believe in those things anymore."

"Oh Charlotte, you know better than that. Papa was a good man, and I know you've seen what the priesthood can do. Remember when I was fourteen and Esther kicked me in the head? Papa called Brother Logan and they gave me a blessing. Well, let me tell you it wasn't that little bit of oil that healed me and gave me back my senses. It was the priesthood, and the faith of those two good men."

She looked at him for a minute in silence, but finally turned away. "Yes, I remember. We all knelt when they prayed, and that time I really prayed. It was for you, and I would have done almost anything for you. But not even for you could I leave my children, no, not even for security and kindness. Jack will mellow. He'll come around someday when he sees he can't break me like he does his horses. Besides, I have already asked about the law, and legally I am his property. I can do nothing without his consent, and if you took me away, you would be breaking the law."

There was silence between them. Finally, she turned around to face her brother. His blue eyes, usually twinkling, were watching her thoughtfully.

"What are you thinking?" she asked.

"There you are, Charlotte O'Neill Boughtman, looking like a speckle-faced, wild, hillbilly girl, a woman of—let's see—about twenty-four years, and you've not changed more than an iota from the girl you were at sixteen."

"What on earth do you mean?" she laughed. "I am a mother and a wife, and I've learned a thing or two I can tell you."

He shook his head, "Not about the things that really count. You are still as stubborn and headstrong as you ever were." He grinned lovingly at his sister, "What'll it take to learn ya?"

Jack had no sooner dismounted than Matthew, all arms and legs, climbed all over him, and Ruby was not far behind. Before the boy told Jack about the frog he had killed or the funeral they had had, he told him about the strange man who came to visit their mother all day and said he was their Uncle Johnny.

"I thought he was quite handsome," Ruby said in all her four year old wisdom.

"I thought he was quite dumb, all citified in his dandy clothes. I bet he don't even know how to ride a horse, much less break one," Matt chimed in.

Jack was already half way to the house. He hit the door without breaking his stride. "Charly!" he yelled. "Charly, where are you?"

At the top of the stairs, he almost ran right into her, coming to meet him. "Who is this uncle Johnny the kids are talking about?" he demanded.

"My brother, John Patrick." She answered him calmly.

"Who the hell told him he could come here to my house, a damned Mormon, isn't he? I know he is. He's the one that was on a 'mission'." Jack's mouth twisted on the word.

"He just came to visit me. And he doesn't need a by-your-leave to come to my house. He's my brother."

"It happens to be my house," Jack said coldly. "I built it and paid for it and all the nice things you take for granted."

"Very well, then, I could meet him in the meadow." She returned him a smile for his glare, and he looked more closely at her. She had washed and pinned up her hair in a neat braided bun on the crown of her head. Her face was shining and creamy white beneath the sun-popped freckles. She had put on fresh clothes, tucked the shirt into the pants and tied them around her waist with the sash from the curtains.

"What are you all dressed up for?" Jack asked suspiciously.

"You can't be serious. All dressed up? In what? I don't have a dress to wear, and the one that Lily made me, you took. What did you do with it? I want it and a few others besides. If you're such a rich man to afford this fine house and furnishings, you can afford a dress or two for your wife."

He stared at her in surprise. For two years she had silently accepted his ways and anything he wanted to do. He had almost gotten used to her being unkempt. He had taken the dress because he wanted to keep her as she was, compliant with his will.

"I bought you a dress once, and you smeared it with tar. Besides, I like you just fine the way you are."

"Very flattering, but not to your odd taste. Most men like their women to look like women."

"What do you know?" he asked sarcastically. "You've always been more of a bobcat than a woman. Pants suit you."

He thought she would shrug her shoulders and walk away. It was her way of avoiding the countless arguments that threatened to arise every day. But instead, she took him by the arm and said, "Come in here for a minute."

In their bedroom, she walked straight to the yellow and white gingham curtains, and with a yank, tore down one side. Then swiftly she shed her boyish outfit, and deftly wrapped the curtain about her, tying it at the waist with the sash. "Does this look like what is supposed to go in pants?" she asked him archly. "I want a fine dress with a neckline low enough to make you remember I am still a woman."

A smile slowly appeared on Jack's face. "Suppose I still say no?"

"Then you'll have the devil's own time keeping curtains around here, Mr. Boughtman."

He reached for his wife, and pulled her up against him. His dark head towered above her own, and she had to tip her head completely back to look him in the eye. "Get out of that curtain, Miss," he said gruffly. They struggled for a moment while she refused to be unwound, until he finally agreed gruffly, "All right, you can have your dresses." Then she kissed him and the curtains were all forgotten.

Jack took Charlotte riding in the meadow that summer. Sometimes, the children were with them; sometimes, they went alone. Jack never told her that he loved her. He never told her he was sorry about her father. But she knew that in his own way he adored her, and his savage act against her family had been in retaliation for her love for them. Jack could not share anything, especially her. Gentleness was not a part of his strange, violent nature, but sometimes she would awaken at night to find him leaning over her, looking down with a wistful expression. Once he asked her, in the dark of the night, as he held her locked in his arms, "Charly, do you still hate me?"

She knew what he meant, and in her pride, hesitated to relent from those painful words. Yet the desire to be happy was strong, and she knew he loved her. At last she said, "I don't want to hate you, so I don't think about . . . things."

He pressed her, "But you don't love me, do you?"

It was a few minutes before she could answer. "I don't know. I don't know if I will ever love you the way I once did."

"But you don't love anyone else?" he asked, still seeking reassurance.

She turned over and faced him, the moonlight lighting just the outline of his rugged face. "Why, Jack Boughtman, who else would I love? You're my husband and plenty of man for me."

He smiled at her, "Just remember it, Miss."

And so they made their peace, Jack content in possessing her, and Charlotte not forgetting, never forgetting, but trying to put it out of her mind enough to live with him.

John Patrick came back a few days after their first visit to say good-bye. Charlotte and Johnny had been talking about an hour, sitting together in the large white swing on the porch, when Jack rode up.

Johnny stood up as Jack dismounted. Charlotte stepped down from the porch and over to her husband, taking his arm.

"Jack, this is my brother, Johnny. He came to say good-bye. He leaves tomorrow for Utah."

Boughtman stood looking at the man on his porch. He looked mightily like the old man O'Neill, and some unwanted memories and emotions flooded him. "Trying to get Charlotte to go with you, I reckon," he said darkly.

Johnny spoke quietly. "I would take her if she wanted to go. But she refused."

Jack looked down at Charlotte by his side. She was wearing the new dress he had just brought her from town. It was pink and white, with a ruffle around the bodice. Tucked into the valley of her bosom, was a delicate pink flower. The heat of the spring day had curled little tendrils of auburn hair about her temples and neck. Pride swelled in him.

"It's a damned good thing for you, too, O'Neill, because you'd never get off the place with my wife. You'd be shot before you were halfway down the lane."

Johnny replied. "If you love her, I can understand why you feel that way. I love her, too. She's my sister, and I leave her here with you only with a lot of regret and," he added pointedly, "concern for her. See that you treat her well, Boughtman."

Jack looked at him narrowly. Johnny's blue eyes returned his dark stare unflinchingly. Johnny came down from the porch and stopped in front of the couple. He spoke to Charlotte as though Jack were invisible.

"Char, I hate to say good-bye again. The last time I did, it was eight years before I saw you again. I hope it won't be that long this time, but I'm afraid it may be." He reached out and chucked her under the chin. "Good-bye, sis. Remember, I'll always love you."

Charlotte left Jack and threw her arms around her brother's neck. "Good-bye, Johnny. Give my love to Mother and Annie." Tears were glistening in her eyes and she searched his face longingly. "Oh, Johnny, come back some day. Will you? Will you come back?"

Neither of them could know how very soon he would be back.

Hector's blacksmith shop was at the west end of Montrose, on the way out of town. It was always hot, summer and winter, and smelled wonderfully (or atrociously, depending on your point of view) of horses, leather, wood, sweat, straw, and molten metal. Hector always had more work than he could do, which didn't concern him in the least. He was a big man, even bigger than John Patrick, with a shock of dark brown hair that frequently hung damply over his forehead and eyes. His hands were scarred and re-scarred from burns, and his wrists and forearms were as big as many men's biceps. Not a man who lived for his work, nevertheless, he was a hard worker, steady and seemingly tireless.

Johnny was sitting on a crate in the doorway where occasionally a breeze would offset the heat of the forge. "Hector, you ever thought of going west?"

"What for?" Hector asked, hammering steadily on the horseshoe.

"They need good blacksmiths on the frontier, and a man could do well for himself."

"I do all right. Don't need no more than I got." Hector didn't look up.

"Haven't you ever wanted to see other places? They say there's Indians with feathers, and buffalo herds and great huge mountains out there. Seems like it would get tiresome staying in one place all your life, seeing the same faces, the same roads, the same everything."

"Nope, don't git tiresome. I'm a homebody." The smithy finished the shoe and set it aside to cool. Then he started on the next one. Wiping his face with his forearm, he asked, "Is that where you're headed?"

"Yep. Going west where a man's got room to breathe, land to settle, and plenty of freedom."

"You ain't got freedom here?"

John chose his words carefully, watching Hector closely to see the impact. "Some . . . some freedom. But a man's gotta be able to worship his God the way he sees fit. Are you a religious man?"

"Not much. I like the singing the most; the preaching I ain't too much for. I don't like to think about the devil and hell fire with an infernal fire burning all the time. I had enough of fires, and one or two burns." He held up the back of his hand.

Johnny nodded his head in understanding. "I don't like that kind of preaching either. I like to hear about the prophets, and when they talked to God."

"Yep, me too. But you don't hear much of that. Mostly what you hear is how wicked and evil you are, and how you got to repent or be burned. I don't like that much."

"My religion talks about that a little. But you hear more about living with God in heaven before we were born on earth, and living with Him again in the resurrection."

Hector was intent on the horseshoe, turning it this way and that in the fire, striking it steadily with his hammer to bend it just the right amount. John wondered if the blacksmith had heard much of what he had said. John felt the stirring of a breeze in the doorway, and flapped his shirt about him to let the cool air into his chest. Perspiration was standing in little beads on the amber hairs of his arm and chest. He looked at Hector and wondered how he could stand over the forge for so long.

"I never hear'd that before."

"What?" Johnny asked.

"Living with God before we was born on earth. How can you live before you are born. It don't make sense." Hector was looking at him now with a puzzled gaze.

"Well, Satan did. The Bible says he was kicked out of heaven and fell to earth, that old serpent. And Jeremiah did. The Lord told him that before Jeremiah was in his mother's belly, God knew him."

"Hadn't ever thought of that."

"I believe we were all God's children in heaven. And then we were sent to earth and got a body. Everybody knows we have a spirit inside the body. But nobody knows where it came from. Well, it came from God, and it's going back to him when we die."

Hector stood quietly, looking at him for several minutes, then turned back to the horseshoe, which was red hot by this time. After a few more minutes, he paused in his work again and pulled his wet shirt over his head. Sweat had drenched the shirt and was rolling in small streams down his back and chest.

Stepping over to a big rain barrel in the corner, he reached in with a ladle and scooped water over his head and shoulders. He stood there, dripping, then turned back to John.

"Where'd you learn sech things?"

"From a man named Joseph Smith."

"Wasn't he that Mormon prophet they killed a few years back?"

"Yep."

"Ain't you ascared he'll lead you to hell?"

"Nope."

"How come?"

"Cause what he preached was true. It makes sense, doesn't it? You ever heard anybody else say where the spirit of man comes from?"

"No, can't say I have . . . But I haven't listened very careful either."

"They don't say, cause they don't know, and they don't know because they aren't prophets. When God has a prophet, He tells things to him, things like where our spirits come from, and why we were sent down here in the first place."

"I'm not shore I believe that God picks prophets any more."

John shrugged carelessly, "Well, that's probably why you don't go to church much. Who *would* believe a man's teachings that wasn't a prophet? Wouldn't do you any good, cause he wouldn't know any more than you did." From under his eyebrows, he watched Hector absorbing all that.

"Want me to work those billows?" John asked.

"If you want something to do, it'd be all right."

So Johnny stripped off his shirt and took his place across from Hector at the forge. They worked together silently for a while. John knew that Hector was mulling over the conversation in his mind. He waited.

"Show me how to do that, will you?" John asked.

"Not much to it. Just keep turning it and hitting it. Sometimes you got to squeeze it together, if you got a horse with a small foot. Got to get it sorta smooth and even as you can. Horse is not so likely to stumble if all his shoes are even."

He handed the tongs and hammer over to John, and watched him as he struck the metal shoe.

"I never heard the Mormons preach before."

Absently John Patrick replied, "Well, you haven't yet, either. For a really good sermon, you ought to hear Brigham Young or Parley Pratt." He kept up the tiring work of gripping heavy tongs and beating metal. It wasn't long before he could feel the strain in his shoulders and forearms.

"Where do they preach?"

Johnny looked up, a little surprised. "Out in the Utah Territory. They've all gone out to Utah. That's way, way out west, somewhere near California. People around here didn't like what they preached and drove them out. So they went west for freedom."

Hector nodded his head knowingly. "I imagine so. People get mighty fussy about religion, and not too understanding about people with different ideas." After a minute he said, "You shore have some strange ideas."

Neither man noticed the small carriage drive up outside, until a portly gentleman in a black coat bustled into the barn.

"Hector, I'm glad to see you here working. I've been here three times looking for my carriage wheel you promised me two months ago, and every time you've been gone fishing or some such thing."

Hector looked at him patiently, "Can't work all the time, Reverend."

"No, but you have to work sometime. The good book tells us there is a time for work as well as a time for play."

"I reckon that's true, and today's my time for work."

The Reverend Henderson tipped up on his toes and settled comfortably back down on his heels. "That's good. Maybe you can get to my carriage wheel today."

"Might be." Hector's attention was unswerving on the horseshoe and hammer that John was wielding. John's back was to the Reverend, and the blacksmith could see his broad smile.

"You have a helper, Hector?" The Reverend walked around the forge so he could get a better look at Johnny.

"Nope, just a customer. But he's a quick learner and he wanted to know how to do it himself."

Johnny looked up from his work, still smiling, despite the perspiration now starting to stream from him also. "How do you do, sir?"

"I do just fine. What's your name, young man?"

"O'Neill, John Patrick O'Neill." John was growing a little tired of the man's imperious nature. "What's your name?"

"Reverend Henderson," the older man stated.

"Papa, is the wheel ready?" A young lady had come several feet through the doorway into the fragrant barn. She was dressed, even on such a warm, spring day, in a heavy, gray gown, buttoned up to her chin and hanging in voluminous folds down to the straw-dirt floor. Her hair, neatly braided and coiled on the back of her head, was the color of ripe wheat.

She was apparently unprepared for the sight of two shirtless men at work, and her eyes flew wide open, while her hand went up to her mouth. Reverend Henderson was disapproving, "Faith, I thought you were going to wait in the buggy."

Sky blue eyes were riveted on Johnny. "I . . . I . . . was going to. But I . . . I thought. I just wanted . . . I'm sorry, Papa."

Johnny's smile broadened, amused by her reaction. Her father moved to her side and started to turn her shoulder away. "Well, he doesn't have it done, just as I said he wouldn't. We might as well go." But Faith wasn't moving.

Johnny handed the hammer and the tongs back to the smithy. "I think we are almost finished with these horseshoes. Maybe we can work on your wheel next." He spoke to her father, but his eyes never left her face.

She was very young, he thought, maybe only fifteen. There was a sweetness to her expression, and an innocence that was completely unfeigned. Incredible blue eyes and creamy white skin that, by now, had begun to turn pink with embarrassment.

She tried to be modest and cast her eyes down, but they immediately fluttered up again, gazing in some wonder on him. Johnny had never felt so self-conscious.

The Reverend forgot his daughter for a moment and said, rather querulously, "Do you really intend to do it now? If so, I'll wait."

Hector replied, "I reckon we could do it now. Mr. O'Neill's horse will have to wait for these shoes to cool down before we can get them on him, anyhow." He lumbered over to the side of the barn and rolled the wheel out. The metal rim around the wood had come loose, and one end had popped away from the wheel entirely.

"Faith, you get back into the buggy. This isn't any place for a girl." Reverend Henderson pointed her in the direction of the door and gave her a little push.

"Yes, Papa. I'm sorry." She tore her gaze away from John Patrick and moved to the barn door, glancing back one last time as she disappeared around the door frame.

Johnny's eyebrows lifted. He let out a little whistling sound and forced his attention back to Hector. The blacksmith had jerked the metal rim off the wheel and had the end of it in the fire, heating the metal so he could work it. John worked the billows and Hector worked the metal. Reverend Henderson stood watching for just a minute. Then wiping the perspiration from his brow and neck with a perfectly white handkerchief, he left with an admonition to keep it up, saying he had an errand at Miss Ann's place. He would be back in a few minutes.

"That's the preacher's daughter," Hector said in a matter-of-fact voice. "Her name's Faith, and she's almost too old to get married. Be twenty years old next birthday, and folks say the Reverend ain't never gonna let her go."

"Oh really! Why's that?" Johnny asked.

"Cause she's his only one. The missus went and died about eight years ago, so he never had no more children. Faith has been a taking care of her pa since she was eleven. He wouldn't know what to do without her."

"I don't blame him. Pretty little thing like that, no wonder he doesn't want to give her away."

"You could be shore he wouldn't never give her away. Man would have to have something special to convince Reverend Henderson to part with his daughter."

"She doesn't look almost twenty. I thought she was about fifteen."

"Nope. I reckon that's cause she's so protected. I imagine that's the first time she ever see'd a man without his shirt."

"Well, I hope it didn't shock her."

Hector looked up from the fire and grinned at Johnny, "She didn't appear to mind it. Leastways, she never took her eyeballs away from you."

Now it was Johnny's turn to grow red with embarrassment. They worked for another half an hour, heating and beating, then cooling and measuring. Finally, the metal rim was as perfect a fit as they could get, and they fitted it back on the carriage wheel. Johnny cinched it up around the wheel, and held it in place while Hector pounded in the thick metal bolts that went through the wood, holding the rim in place.

Before Henderson returned, Johnny had washed up, and put back on his shirt, making himself as presentable as possible. Shortly they heard voices outside the door, Henderson's impatient voice and Faith's softer, sweet voice. Hector motioned to the other side of the barn, "There's a back door over there. I'll take care of the Reverend."

Johnny smiled, gratefully, "Thanks. I'll be back." He climbed over the bales of hay stacked back against the wall and slipped out the door just as Henderson came in.

She was perched up on the seat of the buggy, and Johnny quickly jumped up beside her. Startled, she slipped over to the farthest part of the bench.

"Excuse me, Ma'am, I just thought I'd come out and talk awhile til your Pa is done. Did I hear the Reverend call you 'Faith'?"

She was watching him with curiosity. "Yes, my mother named me that. She said it took a lot of faith for her to get through that pregnancy and more faith to try and raise me."

He smiled at her. "I can't think you were so hard to raise."

"Well, Mama was very sickly, ever since I can remember. I guess I was too much for her. Maybe that's why Papa is so afraid for me."

"Is he? You don't look sickly. Not a bit." She had begun to blush again and looked down at her hands. "In fact, you look the picture of health, with those pretty red cheeks and bright eyes."

"You're embarrassing me, Mr. O'Neill."

"Oh, now, don't mind me. I'm as bad as my Pa. He was always teasing my sisters and Ma, too. I just meant that you're a mighty lovely young lady, Faith Henderson. It's a wonder you're not married."

"Probably never get married."

"Why not."

"Papa needs me. He tells me so all the time. How would he get on without me? He doesn't cook, nor wash clothes, nor iron at all. I couldn't let him down."

"Not even for somebody else you loved?"

She raised her summer eyes to him again briefly. "Never have loved anybody," she whispered.

"It'll happen someday, mark my words. Then what will you do?"

"I don't know."

"Do you live here in Montrose?"

"Just outside of town about three miles."

"Oh."

"But Papa travels all over the county preaching."

"Do you go with him?"

"Yes, mostly I do. He likes me to and I don't like to be home alone."

They were sitting comfortably again and Faith was only a few inches from him. John could smell a light aroma of roses about her, and marveled to himself how smooth her hair and skin were. For eight years, he had been on a mission. He had left Nauvoo at eighteen, still a raw-boned boy, and for eight years had not had time nor inclination to give much thought to girls. Before he had turned eighteen, there had been a few girls he had cast an appreciative eye on. One girl, he had kissed behind the house at a church social one dark night. On his mission there had been girls, pretty enough, and some seemed to be drawn to him; but he moved about quickly and was more concerned with converting their fathers than courting them. It just hadn't been the right time to think about girls. He had learned that such thoughts could start feelings that were hard to deal with, and so preferred to keep his mind strictly on his work. Now those feelings were stirred up and rising, until, as he sat beside the minister's daughter, John wondered what it would be like to hold her in his arms, to feel her head on his shoulder, and her pink lips on his. He tried to keep his attention on what she was saying.

"Where do you live?" she asked.

"Nowhere. I am headed west, looking for a nice spread where I can settle down."

"Oh, I see. So you are just passing through. I thought maybe you were Mr. Thornby's helper in the blacksmithery."

"No. I enjoy learning about everything, though. So I asked him to show me how to work the forge and make a horseshoe. Figured I might need to know such things when I go west. Man's got to be able to help himself."

"I'm sure. What kind of work can you do, Mr. O'Neill. You must do something to earn your bread and butter."

"I do everything, and nothing. I can push a plow, train a dog, break a horse, wield a forge, run a river ferry, build a house, and skin a bear. I expect that I'll find a wife, settle down, and run a successful farm someday. On the other hand, I might be a store owner like my Pa. I could do about anything I set my mind to."

She smiled at his bragging. "I'm sure you could."

"Hey, I got you to smile. You should smile more often. It makes your face look like sunshine, and your eyes like morning glories."

Now her smile widened, and she started to color slightly, "You're teasing me again."

"No," he said seriously. "I'm not teasing at all."

The smile disappeared. Faith drew in a breath and held it, looking at him wide-eyed. John Patrick stared back at her, and if he had not been so before, he was thoroughly smitten at that moment. He might have reached out and touched that tender face, but the Reverend's voice broke their spell.

"Ready to go, daughter? Thornby's finally got my wheel done, and I . . ." He stopped short, seeing John Patrick up in the buggy beside his Faith.

John climbed down. "I thought you were gone, young man," the Reverend said darkly.

"Just took a walk, sir, and ended up talking to your daughter. You are to be complimented on such a fine job of raising a young girl. She is a credit to you."

Henderson loved nothing so much as praise. "Well, thank you. Faith is my joy. That's what I always tell her. Couldn't get along without her. She's a good girl, that's for sure."

Faith was looking determinedly down. John handed the Reverend his reins and patted the horse's rump. "Very fine meeting you, sir. And you, Miss Faith."

But she would not look at him again, and the buggy pulled smartly away.

Hector and Johnny walked into Miss Ann's dining room at dinner time and sat down. They had both cleaned up and put on a change of clothes. Miss Ann was not a Miss at all, but a widowed woman who ran a bed-and-board house in Montrose for occasional travelers on their way to Chicago. It was a place where the single men could get a good meal. Tonight, there were three guests at her table, Hector Thornby, John Patrick and a fancy-dressed gentleman who looked to Johnny like a river-boat gambler.

Over Miss Ann's biscuits and mashed potatoes, Wade Walters told them his story. He was a gambler, it turned out, and down on his luck. He was coming from Chicago headed south to St. Louis. He had lost his shirt in Chicago to some big time gamblers and had lost his wife at the same time. Walters had married the seventeen year old daughter of his father's best friend. This old, family friend had died shortly thereafter, leaving his daughter with his only earthly possession—a twenty acre piece of land near Montrose, Iowa. He had left Walters with an admonition to take good care of his daughter, and Walters had tried. He actually loved her—so he swore to his two listeners—but gambling was in his blood, and he lost all their money satisfying the craving. They moved from one apartment to another to avoid landlords and bill collectors. Finally, one night a few months ago, his young bride came down with scarlet fever and died. Now he had nothing left in life but this piece of land just outside town.

"Are you going to settle down then and work the land?" Johnny asked curiously.

"I suppose I should," Walters answered, his thin hand wandering nervously about the tablecloth. "I hate farming, though. I'm really a city boy. But I guess I owe it to the memory of my wife to try and change my ways."

"You mentioned you were heading to St. Louis. What will you do there?"

"Get a steady job, try and get a little money ahead, then come back here and build a house and try my hand at farming." Walters pushed back his chair. He was rapier thin, and a little pale, dressed in the threadbare clothes of a dandy used to better days.

"I'm really a pretty good gambler. If I could get into a game, I could win myself a stake in no time. But I can't even raise the money to join a proper game."

"How much do you need?" John asked.

"Oh, about fifty dollars." Walters was watching him intently.

"Don't you have anything worth fifty dollars?"

"Sure, a horse and saddle, but I can't sell that."

"How much do you figure your land is worth?"

Walters began unconsciously to lick his lips, and his hand nervously smoothed down his hair. "I reckon it'd be worth about a hundred, but I couldn't sell that, either."

"Of course not. I just wondered." John went back to his bread pudding with relish. Hector glanced at him quizzically.

"On the other hand, if I can't get no money to build on the land, it won't do me no good." Walters was more and more nervous, fidgeting in his chair, scratching his left ear.

John waited and ate.

"You interested in farming?" the gambler asked John.

"Me?" John asked in surprise. "Not much."

"Maybe you had something else in mind for the land?"

"Oh, I don't need land. I'm moving west."

Now Walters was puzzled. "Well, what was you asking me all those questions for about the land and all? I thought you was interested in it."

"You interested in selling?"

"If I could get the right price. One hundred dollars, that's a good price for the land."

"Well, that settles it. I haven't got one hundred dollars, anyway."

"How much you got?"

"None, really." Wade Walters' face fell. "But I do have something you might be able to get some money for."

"What's that?" Walters asked again, licking his lips once more.

Johnny slowly fished around in his shirt pocket and pulled out a gold piece. "Just this. It's a Spanish dubloon. Had a man tell me once it was worth more than a hundred dollars."

Walters reached out for it, greedily. "Spanish gold!" He stuck it between his teeth. "Looks old, all right. Where'd you get it?"

"A run-away slave gave it to me once for helping him. Where he got it, I don't know."

Wade turned it over in his hand several times, looking at the markings on it—Spanish words he couldn't understand, and the numerals 1634 that he could. "You interested in the land? I'll trade you. This gold piece is worth maybe fifty dollars, but I'll trade you."

"I wouldn't want to cheat you, Mr. Walters. Besides, I'm not a farmer either." Hector was puzzled, wondering what Johnny was up to. But John had decided, when Reverend Henderson had driven away, that he somehow had to stay in Montrose. Besides Faith, he reasoned, he owed it to his sister Charlotte not to leave without trying to convert her.

"It's a mighty good piece of land, mister," the gambler said, reluctant to give the gold piece back.

"Well, that dubloon is kind of a keepsake. I'm not sure I want to part with it."

"I'd consider a trade." Walters persisted.

"Only if you're sure the gold piece is really fair compensation." Johnny had pushed back his chair and was enjoying the exchange.

"You said it was worth a hundred. You look like an honest man. I'll believe you and take a chance. Is it a bargain?"

"You sure your land is good land? I don't want any river bottoms!"

"Oh, now! It ain't no river bottom land. It's good farm land, just out of town to the east there about three or four miles."

"Any buildings on it?"

"No, no buildings, but an old chimney you could build a house around, if you had a mind to."

"Hector, can you feature the property he means?" Johnny turned to his new-found friend.

"I think so. Like he said, it's out of town to the east. Fair land. Not far from Henderson's place." Hector stared soberly into Johnny's twinkling blue eyes.

"My friend here vouches for you. It's a bargain."

The gambler jumped up, stuck his hand out. "Shake on it?"

"Sure," smiled John. And they shook. Walters snatched up the gold piece and stuck it in his vest. Then he pulled out a piece of wrinkled up paper. "Here's the deed. It's all legal and signed like it is supposed to be."

"Good," John said. "Now, you sign the back. That shows you traded it to me."

The gambler signed and John became a neighbor to Faith Henderson.

CHAPTER 6

Three weeks of sawing and plowing and clearing from daybreak to dark had hardened Johnny's muscles and trimmed almost ten pounds from his already fit young body. The May sun had bleached gold into his red head and tanned his freckled face. He had succeeded in clearing a good acre of land round about the old chimney. He had it marked off, a spot for his house, a little garden, and space for a barn. He had cleared the garden area first and planted it, as soon as he had the stones removed and the land plowed. He knew he was late planting but hoped that with good weather he could still get a crop, enough to live on anyway. One morning he had opened his Bible for a few minutes of reading. It had fallen open to Mark, Chapter 4:40, and he read, "How is it that ye have no faith?" Johnny threw back his head and roared with laughter, dashed outside and saddled his horse, and off he rode to the Henderson place.

Faith was outside behind the house, hanging clothes on a clothesline stretched between two trees. She was obviously not expecting company and had not yet braided and done up her hair. She was dressed in a simple house dress with no voluminous petticoats to hide her slight young figure. John sat on his horse, watching her for several minutes. She was bending and stretching to hang up the clothes, while the wind was playing havoc with her hair and blowing her skirt around. She was, he thought, altogether lovely. Then kneeing his horse, he trotted up to the girl.

Faith spun around, "What on earth? . . . I thought you were long gone out west."

"Did you. Guess you were wrong!" He grinned impishly at her.

"What are you doing here? Are you here to see Papa?"

"Not so's you'd notice it." John's eyes were twinkling as he enjoyed her confusion.

"Well . . . what are you here for then?"

"Just taking a ride and came upon a young lady 'faithfully' performing her duties."

"Now stop teasing me about my name."

"Wouldn't do it, ma'am. Absolutely wouldn't tease you about such a fine name. In fact, you know I came across your name in the scriptures just this morning. God asked me right through His holy scriptures and the Apostle Mark, 'How is it that ye have no Faith?' And then again He said, 'Where is your Faith?' And I said to him, 'I truly don't know'. And He said to me, 'John Patrick O'Neill, get yourself out and find Faith!' So here I am."

By this time, she was laughing, her hand over her mouth to control it. He slid down from his horse. "And now I have found it—my Faith, that is—and the next time the Lord asks me that question, I can say, 'right over yonder, that's where my Faith is. Right over there, hanging out clothes.' Now, who would have thought Faith would be found hanging out clothes."

He took her by the shoulders, both of them laughing at his silliness. He was almost a head taller than she, and she had to tip her head back to look at him. In the sunlight, with her blonde hair blowing around them, Johnny's heart swelled, and he bent down to touch those innocent lips. But she pulled away, suddenly sober.

"You have such a way about you, Mr. O'Neill. You get me all . . ."

"Not Mr. O'Neill. My real name is John Patrick, the other is just an alias."

"What's that, an alias?"

"Oh, just something I tell people."

"You're crazy. How'd you get to be so foolish? The scriptures warn against people being too light-minded."

"You reckon so? Doesn't the Bible say the fruit of the spirit is love, joy and peace?"

"Yes, but . . ."

"How can you be joyful, and show the Lord you're happy with a long face?"

"I . . . I don't know, but you seem . . . you confuse me," she said finally.

"That's nice. I like that."

"You do? Why?"

"Cause you're so pretty when you're confused."

With some effort, she pulled away from his gaze and bent to take another shirt from her wet pile of laundry.

"Aren't you going to ask me what I am doing hanging around these parts?" Johnny watched her appreciatively.

"All right, what are you doing hanging around these parts?" She reached up to the clothesline. He put his hand over the shirt to hold it on the line until she could pin it.

"Helping you with the laundry."

"Aren't you ever serious?" she looked at him in exasperation.

The laughter faded from his eyes, and he looked down into her face. "Sometimes. I bought a piece of land a couple of miles away. I just spent the last three weeks clearing it. I have a garden planted, and tomorrow I start my house."

Her mouth dropped open. "What are you talking about? You were heading out west."

"That's right. But I met a gambler who wanted to get rid of this piece of property, so I took it off his hands. And here I am." His other hand closed on hers over the clothesline. Her face was a canvas of fleeting emotions, surprise, disbelief, joy, discipline, confusion. He thought he would never tire of watching it.

"You're . . . you're settling down here?"

"Uh huh. Two miles away. Just through those trees at the edge of the meadow. Not far. Just whistling distance, actually. Can you whistle, Faith Henderson?"

She pursed her lips obediently, just inches from his. Her eyes caught his just as he bent to kiss her, but she jerked her hand away and turned her back to him. He heard her whisper something.

"What did you say?" he asked.

"I said," she turned half way back. "I said you upset me. I'm not at all sure I will like having you for a neighbor." Then she looked him straight in the eyes, and her chin came up defiantly, "I'm not at all sure I like you."

"Why," Johnny asked seriously.

"Because you're so . . . so clever, and so flippant, and very fresh. Other men don't treat me this way. They respect my father, and they respect me as his daughter. You're a strange man. You can quote the Bible, but you don't seem to be religious."

"Why do you say that?"

"Because you are always laughing about something. You're not sober enough."

"Perhaps my 'Faith' has brought me joy."

"There you go again," she stamped her foot impatiently. "Making jokes about my name, and it is a perfectly good name."

"Oh, it is! I certainly agree. And Faith is a wonderful thing. I'd really like to have more Faith." Johnny couldn't help himself. She was so delightful to tease, and it was so easy to do with a name like hers. But now she began to get angry, and gathering up her basket in a huff, started for the house. He ran after her, grabbed her arm and started to apologize, when the door opened and her father came out.

"Faith," the Reverend hollered, "who's this stranger?" Then he stopped, recognizing Johnny. Reverend Henderson was in his shirt sleeves, with his suspenders swelling over his belly. Well fed, thought John. She must be a good cook.

"Thought I'd stop by and say hello to my neighbors," Johnny ventured before the old man could ask.

The Reverend frowned. Johnny took the basket of clothes from Faith. "Here, I just wanted to help with your basket." Faith looked away and mumbled a 'thank you'. John placed it back over by the clothesline, and continued to talk to her father.

"Yep, bought a piece of land not far from here. Folks around told me how lucky I am to have a fine neighbor like Reverend Henderson. Always got a warm fireside for a stranger, and a hand of welcome. That's a fine reputation to have, sir."

Henderson shook the hand now extended to him. "Well, thank you, young man. I try to be friendly, especially to people I think deserving." He started to lead the way back inside. "Are you a God-fearing man?"

"Well, I think so, Reverend. My folks taught me from the scriptures, and I sure do like a good sermon. Are you a sermonizer?"

"Certainly. Love to preach the Bible. I travel all over this country preaching and performing weddings, sometimes preaching a funeral. It's the Lord's work and it's mighty satisfying, I can tell you that. Here, set a while and we'll talk. Faith," he called, "Get some lemonade for us men."

She came quickly in the door and hurried to the kitchen. John Patrick could see it was a nice house. Besides the parlor he was in, there was a kitchen, large enough for a dining table, and off of the kitchen he could see two bedroom doors. The front of the house had a wide veranda for sitting and rocking. Faith reappeared in moments with tall glasses of ice-cold lemonade. She served them silently. Then, with eyes still averted from John, she went back to her clothes-hanging task.

"Where's your land, O'Neill?"

"About two miles north of here, right through the clump of trees at the far end of this meadow."

"Gonna farm it?"

"Yes, sir. I've got an acre cleared, and most of it plowed and planted."

"Late planting."

"I know. But I'll get enough to keep me in food this year. Maybe next year, I can produce more and sell some."

"You gonna be a regular church attender?"

"I can't really say, Reverend. Where is your church?"

"Anywhere we feel like setting it up. I preach from a tent. Folks around here are really close with their money. Haven't got a lot anyhow, and I never have been able to get them to build the town a church. Sometimes we meet at Miss Ann's, if the weather's poor. Most of the time, in the summer, we set up the tent, and people come and enjoy the evening air."

"Well, what kind of doctrine do you preach?" Johnny asked innocently.

"The Bible, son, I just preach the Bible."

"I understand, sir, Christian charity, love thy neighbor, and the writings of the prophets and all that."

"That's right. I get right wound up about the Sermon on the Mount, and the Ten Commandments. People have a tendency to forget those good old- fashioned commandments."

"I agree. They sure do. Why, I've known some good church-going, pious men who would walk right out of Sunday meeting, go home and beat their wives and daughters, and cheat their neighbor out of his best hog."

"It's a shame all right."

"It's just as though they think the good Lord can't look right down and see all the mean things they're doing." Johnny paused, shaking his head. "Do you think he really can, sir? See us from way up there in Heaven, I mean?"

"Course he can! Course he can. God sees everything."

"I believe it! Do you suppose he can hear us, too?"

"For a fact! He sure can."

"Have you ever heard the voice of the Lord, Reverend?" Johnny was leaning toward Henderson, hanging on the man's every word.

"Well, I can't say as I have heard a real voice, not like you and me a- talkin'. But I must say I've heard him whisper things to me many a time. Just a real feeling, like he is whispering in my heart."

"Really? Well, maybe He doesn't have a voice. Maybe God can't actually talk?"

"Of course he can." The Reverend snorted. "Moses talked with him face to face. Jonah heard the voice of the Lord. Abraham heard it."

"And you've heard it. Boy, that's really something! I'm proud to meet such a man as yourself, Reverend. It must be a special feeling to stand alongside of those prophets of old."

"Ah pshaw, I ain't to be compared to those men. I'm not a real prophet or anything. God doesn't call such men in this day."

"Now, Reverend Henderson, I don't believe that. You're probably as good a man as there is, and you just said how the Lord talks to you sometimes. I figure that makes you pretty close to being a prophet too. And, truth is, I wouldn't ever follow a man that wasn't a prophet. Wouldn't be any use. You know, you've answered a lot of questions for me. I used to wonder if God actually had a body of parts like you and me. Now you've cleared that up."

Reverend Henderson was beginning to frown into his lemonade, and shift about in his chair. "What do you mean?"

"Well, if He can see, He must have eyes. If He can hear, He must have ears. He couldn't talk if He didn't have a voice and a mouth, and certainly Moses couldn't see Him face to face if God didn't have a face. Now that is a true doctrine to me, sir—that our Father and God has a body just like we do."

"Whoa down, there, son. I never said those things."

"You sure did. And it makes a lot of sense, too."

"What religion are you, young man?"

Johnny stood up. "I'm a Mormon, sir. And I've never heard anybody explain the reality of God better than you have. Maybe I'll be to your tent meeting. When is it?"

"Friday night. A Mormon, huh. I thought they had all left these parts years ago."

Johnny smiled and started for the door. "Most of them did. I'm just a laggard."

"Well, you'd better be careful not to spread that around town. Some people don't like Mormons even yet. And I'm not sure I ain't one of them."

"Thanks for the advice and the talk. I enjoyed talking doctrine with you. You're a fine preacher." Johnny tipped his hat to Faith who was standing by the clothesline watching him with her father, and when he smiled at her just before circling his horse and riding off, she dropped her Papa's best white shirt in the dirt. "Oh, drat it all," she exclaimed as vehemently as she could. "Now I'll have to wash it all over again."

Friday evening was warm and balmy. On the eastern outskirts of town, not far off the road, Reverend Henderson and a few faithful followers had set up a tent. Outside of it they had a potpourri of chairs that were soon occupied with many of the good people of Montrose. Johnny sat on his horse in the deepening shadows of the trees and surveyed the scene. He wondered how many of those good men, sitting and listening to the Reverend's haranguing voice, had been the ones to carry the tar buckets and the fire brands when they had burned out his father's homestead. For a few minutes sitting there, he wrestled with angry, indignant thoughts. He had thought nothing in the world could have gotten him to stay in Montrose because of that event. He had wondered how Charlotte ever stomached the town. And yet, here he was, building a house and clearing land outside of Montrose. His eye searched the backs of many heads, looking for a neatly braided coil of wheat-colored hair. At last he saw her, sitting on the front row, squashed in between two ample women. She was a light little figure, slender as a willow wand, and Johnny wished he could see her face. Now and then she glanced around, as if looking for someone. But he was securely hidden behind some low hanging branches of a maple tree and the grass that grew almost shoulder high.

Perhaps that is why another man, also standing in the shadows, didn't see him. Jack Boughtman came slipping through the tall grass. At first, John wondered if his eyes weren't playing tricks on him. But no; it was Boughtman. Johnny watched him curiously. What was he doing at a revival meeting, Johnny wondered? Then he saw Boughtman break out of the grass and creep up behind a young, dark-haired girl sitting on the last row. He tapped her on the shoulder, and she turned around. She was pretty enough, but dressed in a tight-fitting, low necked dress, which she tried to keep covered with a shawl. Sitting beside her was a young boy, perhaps eight. Boughtman talked to her a few minutes, jerked his head in the opposite direction, and she shook her head. Jack eased back into the concealing shadow of a tree, just on the edge of the clearing.

Through Reverend Henderson's whining, cajoling, rebuking sermon, Boughtman leaned against a tree and whittled on a piece of wood. When the thing was blessedly over, and the last strains of "Shall We Gather At The River" had died away, Johnny saw Boughtman emerge once more and catch the girl's arm. She said something to the little boy, then inconspicuously slipped away into the shadows with Jack. Johnny mulled it over in his mind. Charlotte must not know. She who was so adamant against polygamy—demanding that her husband have only one woman in his life—she would be so completely undone if she knew

Jack was stepping out on her. What should he do? Should he warn her of her husband's unfaithfulness? Maybe she already knew and he would only shame her. Maybe she didn't. Would it force her into doing something drastic? Would it throw her into the arms of the Church? He shook his head, doubting it. It would, he decided, only embitter her. No, he would not be the one to tell her, at least not now. Maybe after he had had a chance to talk to her more. So far, he had had only one more visit with Charlotte. After leaving Henderson's, he had ridden three more miles east and found his sister at home, sewing curtains. He had told her about his purchase of land, and she had been overjoyed.

"Oh, Johnny, are you really going to stay with me? Are you really staying, and building a farm. It sounds like you're settling down. I thought you were determined to go to Utah."

"I was. But I found a few things around here that interest me. And I met a gambler who traded me twenty acres for a gold piece. So I took it. I figure I can always sell it for more after I clear it and build a house. Besides, it will give us a chance to get to know each other again."

Charlotte wanted him to stay all day, but when he wouldn't, begging off on account of his work, she vowed to come see his place and bring him lunch one day soon. John smiled, thinking of his sister and her obvious delight in him. He found himself hoping that he could melt down some of Faith Henderson's prim and proper reserve and see the same delight in her eyes.

Right now, Faith was walking toward the last row of chairs, saying good night to people as they left. He waited for his moment. The Reverend was surrounded by a clutch of babbling women, talking about his wonderful sermon. Faith's last companion had hastened off, and she was standing with her back to him. Instantly, he stepped out of the shadows and clapped his hand on her mouth, at the same time dragging her backward into the grass. She struggled in his arms, squirming and stamping her feet, clawing at his arm with her hands. Then he could feel her terror, and was immediately sorry.

"Shh," he whispered. "Shh, I won't hurt you. It's just me, Johnny. I'm sorry I scared you, but I'm not going to hurt you. Promise you won't scream."

She shook her head, her eyes wide with fright and anger.

"What's that mean? No, you won't scream or no, you won't cooperate?"

She stamped hard on his foot. "Oww! You are a feisty little thing. Now settle down. I'm gonna turn you loose, but don't run off. I just wanted to talk to you. Please, Faith." Slowly he loosened his hand on her mouth. In a flash, she turned and slapped him as hard as she could, her eyes darting lightning at him.

He grabbed her hand and held it as she struggled. "Faith, I'm sorry. I shouldn't have scared you like that. I just wanted to talk to you alone, and I couldn't figure how to get you away from all those people."

"Well, you can't talk to me in the bushes! I don't go off into the bushes with anyone, for any reason. You are a terrible man. Now I know I don't like you!" He slipped an arm around her waist, and put his hand back over her mouth, pulling her close to him.

"I'm sure sorry you're so determined not to like me. It's mighty hard to love somebody you don't like, and it's even harder to marry someone you don't like." Blue eyes widened as she stopped struggling. Slowly, he drew his hand from her mouth, dragging his fingertips gently across her lips as he did. Then he cupped her face in his hands, searching her eyes. The fright and anger had gone, replaced by wonder. She stood perfectly still, like a doe, watching and listening, and prepared to run if startled. Carefully he kissed her silky hair, then her forehead, and when she made no protest, Johnny dipped his head and touched those lips with his.

Had the moon been put out? Had the world gone away somewhere? What happened to the trees and the grass and the sounds of the revival meeting folding up? Was her father calling her? Neither of them heard or knew anything else but that sweet moment, their lips barely touching, breathing together, bodies gently touching. Then he grew fierce and crushed her to him, trembling with the effort of loving so intensely.

"Oh, Johnny," she whispered in wonder, awed at this tender, beautiful feeling.

"My sweet love. My sweet, sweet love. Marry me, Faith, tomorrow, tonight. By all that is holy to me, I love you. I adore you."

"Yes, of course." It was so simple. There was nothing else, no father to disapprove, no house to build. They were young and in love, and nothing could stay the flow of it.

He kissed her again, and this time her lips were willing as she leaned into him, pressing herself against him. Perfectly trusting, perfectly innocent, she put her arms about his neck and gave her kiss, her heart, willingly to him.

He pulled away first, looking down at her. So full of love and tenderness, he said, "Yes, of course? Just like that, you'll marry a man you already know you don't like? Why, Faith, what kind of sense does that make?"

She smiled charmingly up at him, arms still encircling his neck. "None. It doesn't make any sense. But neither does this thing I'm feeling. I've only met you twice in my life. But I knew from the first you had some kind of power over me. Now I know. It's the power of love. Kiss me again, Johnny. Can we be married tomorrow?"

Shaking his head in disbelief, he gazed at her in wonder. Then he kissed her forehead, her closed eyes, her hair, her soft skin and finally her two sweet lips. Mumbling against her mouth, he said, "Your father is calling. You'd better get back before he has a chance to really come looking."

"How will we tell him?"

"I don't know. I'll think of something." She tore herself away from his arms. Breathing deeply to regain some composure, she moved away. As she turned to go back into the clearing, John Patrick grabbed her hand once more. "Remember," he whispered, "I love you and I'm going to marry you, no matter what. You said yes once. You can't go back on it."

Smiling coquettishly at him, she whispered back, "I won't." Then she darted back into the clearing, and began bringing chairs up to the tent.

"And where were you, Faith?" The Reverend was thoroughly displeased.

"Rounding up chairs, Papa."

John Patrick stood watching her, every look carried a caress he longed to give her. Every so often, she would glance back to their spot, knowing he was still there watching her every move. So this is love, he thought, it's nice!

The next week was June, and John Patrick was industriously working on a house. Now he had an urgent reason to put it up as quickly as possible. He would marry Faith as soon as he had a place to bring her. Hector had traveled out to O'Neill's place a few times to help him in felling trees and splitting logs. He had grown to like the Mormon fellow a lot, and they frequently talked religion. Johnny was wise. He never urged it on Hector. He just answered questions, and then backed off to let his friend mull it over in his mind.

John was splitting logs, his shirt off, shoulders and chest streaked with dirt and sweat, chips and sawdust all over the ground at his feet, when Charlotte rode in on Shannon. He looked up and waved as she came through the path in the trees. Her hair was braided in one long pigtail down her back, and she was still dressed in trousers, but they were becoming, and there was a long sash tied at her waist. Johnny watched her and smiled in delight. She slid down off the horse, tossed the reins over his neck, and started to throw her arms around him when she realized how dirty he was.

"Ick, you look like a hog that's been rolling round in the dirt," she teased as she drew away in mock horror.

"That's a fine thing to say about your darling brother. I'm working! A man's got to get a little dirty when he's working. Besides, company coming unexpected has to take what they find. What 'cha doing?"

Just out for a ride and wanted to find your place." She looked around approvingly. "You've really got a lot done in a month, Johnny. Your garden's coming up, and it looks like your house is, too."

He had the logs stacked up to his waist, four walls around. It would be a good-sized structure when he finished—a bedroom, a parlor, a kitchen, a pantry, and wide front porch. It felt to him like slow work, but each night, he had raised the outside wall perceptively. Of course, once the logs were in place, he then had fill-in work to do with clay, so there would be no chinks and spaces for the wind and rain to blow in.

"If Pa were here, we could have a real good time building this house," Johnny said, leaning on his axe. "I remember when we built the old place in Nauvoo. I was just little then, twelve years old or so, but Pa made me think I was doing the work of a man, and I really worked hard for him. He used to sing. Remember how he used to sing those funny old Irish songs?"

"I remember," Charlotte said. She watched her brother, the tender smile that played across his eyes and mouth, and saw their father again. Memories came back that she had put out of her mind. Memories that were too dear and hurt too much. Somehow, today, they didn't hurt quite so much. Shared with John Patrick, they were more comfortable.

"It's a good thing Pa taught me how to do it 'cause now I've got my own house to build." John looked sideways at Charlotte. "And my own family to build it for."

"What? What are you talking about?" Her mouth had dropped in surprise.

"I'm getting married, Char. I'm gonna marry Faith Henderson just as soon as I get this house done."

"Henderson! Reverend Henderson's daughter? He'll never let you. You're a Mormon. Does he know it?"

"What, that I'm a Mormon? Yes. I already told him. Or that I'm gonna marry his daughter? No. Haven't told him that yet. Just told her three nights ago."

"And she said yes?"

"Uh huh." Johnny grinned and shook his head. "Boy, did she say yes. You find a little girl that's never been in love before, and they really like the experience."

"I suppose it's old hat to you," she grinned right back.

"Nope. I have to admit, I'm just as bad as she is—completely gone, up here," and he touched his head.

Charlotte felt a deep heaviness and disappointment, and at the same time, a joy for this brother she loved so dearly. She had had him back again for such a brief time. It would be different when he was married. Maybe that was how he had felt about her, too. That's the way life is, she thought, so happy so briefly, then change comes, and those happy moments are tiny stars of light in a dark night.

They stood talking, Johnny telling her about Faith and how they met. His arm was draped across her shoulders, and she alternately looked at the ground and at his face, so alive with happiness. They were in that position when Faith drove the buggy into the clearing. She had pleaded with her father to let her go alone into town to get a few things for her kitchen, and to visit a friend, Miss Ann's daughter. She had dressed in her summer frock of white, with pink and blue ribbons worked into the hemline and the neck. Perched on her golden head was a white straw hat to shield her skin from the sun. Of course, she really would go into town, but her first stop was at John Patrick O'Neill's. The memory of Friday night's kiss was still on her lips, and she longed to see him again. She couldn't believe she had so easily promised to marry him. But there was no question in her mind about her decision. She was nineteen, past marriageable age in many circles. But she had been so sheltered she had never even kissed another man besides Johnny. Her father was so modest that she had never seen him in anything but full dress, trousers and shirt. When she had walked into the blacksmith shop and seen Johnny stripped to his waist, glistening and strong, her breath had stopped. He was perfectly beautiful to her. He was also wonderfully entertaining, although a little frustrating with all his teasing. She had known from their first talk in the buggy outside the blacksmith's shop that she was irresistibly drawn to him. Then the day she was hanging out clothes, she had all but forgotten where they were, and had almost allowed him to kiss her then. Friday night was the most beautiful dream of her life, and since that time, she had lived in a dream world with him, until her father had begun to believe that she had taken sick. So the Reverend was very happy to see her perk up on Monday and want to drive into town.

Her buggy came dashing into the clearing, her horse at a fast trot, and she had pulled up a few feet away from Johnny and Charlotte before she could assimilate the scene before her. There he stood, her first love, with his arm around another woman. Tears stung her eyes, sharp pain replacing the joy that had just a moment ago filled her heart. Her breath caught, and she grabbed the whip, flicked the horse's rump, and clucked for him to go.

John Patrick caught it all in a flash—Faith coming to be alone with him, and finding him with Charlotte, whom she did not know. He saw the hurt flash on her face, and as she grabbed at the whip, he broke away from his sister, leaping for the buggy. It was already moving when he jumped in. The horse was almost to the tree line when he got it reined in.

Faith looked at him, her eyes swimming in tears. "Get out of my buggy and go on back to whatever you were doing with her."

"Well now, maybe I will. We were having a wonderful time reminiscing about our Pa together. Faith, sweetheart, that is my sister Charlotte. Charlotte O'Neill Boughtman."

She quickly wiped the tears from both eyes. "Your sister? Charlotte Boughtman? Isn't she Jack Boughtman's wife? I've heard a lot of things about her but I've never seen her. She's a recluse they say. She never sees anybody."

Johnny said gently, "That's because she's so unhappy. Boughtman is a terrible man. He was one of the mob responsible for killing our Pa."

"Why did he do that?"

"Because we were Mormons, and they drove our folks away. They burned them out, tarred and feathered them, and then when they were gone, stole their land."

Faith's eyes were glued to Johnny's face. She asked in a tiny voice, "Are you a Mormon, too?"

"Yes," he said holding her gaze. "Does that make a difference?"

She only paused a fraction of a moment. "Only that it'll make it harder with Papa."

"You sure you could love a Mormon?"

"No! But I could love you, Johnny, and I do, whether you're a Mormon or a Baptist or a Papist or nothing at all."

He reached out to touch her hair and cheek. "I'm too dirty for you to even see, much less to kiss."

"Your lips aren't dirty, are they?" she asked, smiling shyly at him.

"If they are, I'll just lick them and they won't be."

So he did, then leaned forward barely touching her lips with his. After a moment, neither of them cared whether he was dirty or not, and he took her in his arms.

Charlotte waited curiously. Finally, he turned the buggy around and drove back. He helped Faith out of the buggy and introduced her to Charlotte. Faith was shy and thought his sister was a very pretty woman, but strange. What was she doing dressed in men's trousers? She looked up and could see Johnny was obviously fond of his sister. Charlotte was curious and thought that Johnny's fiance was a very young girl, obviously naive but very feminine. *I guess I never have been helpless and soft and feminine*, she thought ruefully. Then defiance came again. *But she wouldn't last a day with Jack Boughtman. He would eat her for breakfast.*

Faith had brought a lunch basket of food. Johnny went to clean up a bit and put on a shirt, and the two women stood admiring the house. Faith's heart swelled as she saw the hard work he was doing and knew it was for her. Peace settled over her, and she knew without a doubt, that she would marry this redheaded, teasing man. They had lunch underneath the trees, and Faith gradually became more comfortable in Charlotte's presence, seeing the similarities between brother and sister. She saw them tease each other and both laugh the same way, with heads tossed back. But she also saw a depth of sadness in Charlotte's eyes that was not there in Johnny's. Finally Charlotte said good-bye, mounted her chestnut horse and rode off. Faith stayed on for a while. She and John walked around the house, and he described it the way it would be when he had it all finished.

"I'd better go to work making rugs and tablecloths and such," she said.

"I guess so. I'll build it, and you make it home," Johnny said hugging her close.

"Have you decided how to tell Papa about us?" she asked.

"No, haven't decided the best way yet. I could get him to marry us without even knowing it, I think." Johnny's eyes were twinkling now in merriment. "I was sort of thinking we could all three be talking, holding on to the Bible, and you could say how you wanted to marry me, and I could say how I wanted to marry you. Then I'd say, 'Reverend Henderson, sir, what is that you're supposed to say when two people get married?' And he'd say 'I now pronounce you man and wife'. And I'd say 'thank you, sir, very kindly. I'll take her.' "

Faith was laughing and shaking him. "Oh Johnny, you are so foolish I can't stand it. I believe you could probably do it, too."

Solemnly he said, "I have certain 'Faith' that I could."

"Am I going to have to live all my life with bad jokes about my name?"

"I'll tell you what. For every bad joke you have to put up with, you'll get something good to offset it." He cocked his eyebrow wickedly.

"What?" she asked breathlessly.

And he kissed her long and soundly.

With Hector's help, Johnny had the house up, roof on, and most of the fill- in work done by the middle of July. He worked day and night. Love was driving him harder than the hounds of hell could have. He could hardly see Faith now without loving her so much he longed to make her his wife. Faith wanted to be close to him every minute, her head on his shoulder, her hand on his arm, rubbing his back, or caressing his hair. Love was new and wonderful to her, and Johnny was amazed that such a beautiful experience had been waiting for him, and he had never realized it.

They still had not decided how to broach the subject of marriage with her father. As it turned out, they didn't have to. Reverend Henderson, for all his pompous ways, was not a fool. He had eyes. He noticed his daughter blooming. Her cheeks were frequently pink, especially around that O'Neill fellow. He began to count how often they ran into Johnny, and how determinedly Faith kept her eyes on the ground, instead of looking at the young man. Once or twice her father had seen her face when she did look at John, and saw a new beauty shining there.

One night in early July, dinner was over and the Reverend and Faith were sitting on the porch, rocking and watching the road. A bay horse came trotting down the road. Faith sat forward in her chair, gazing expectantly at the horse and rider. Her face was glowing, and when her father spoke, she glanced at him, her eyes shining.

"Daughter, I don't believe that's Mr. O'Neill. He was just by here last night."

She sat back in the chair, not looking at her father now, but studiously working on a piece of lace she was tatting. He waited, but she didn't speak.

"It seems to me we see John O'Neill quite frequently lately. He seems to be everywhere . . . everywhere but at tent meeting."

"Oh, he's very God-fearing, Papa. I'm quite sure he is."

"He's a Mormon."

"Is he?" she said. "Did he tell you so?"

"Yes, he did, on one of his very first visits here. I should have told him right then not to come back, but I was too kind. That's my own failing, being kind to a fault."

"Yes, Papa, you certainly are an example of Christian charity."

"But, Mormon or not, I like Mr. O'Neill, and I would like to continue to enjoy his company," the Reverend said pointedly. "I'm having a hard time, however, as I notice that you also seem to enjoy his company *in the extreme*."

Meekly she answered, "I just try to be nice."

"Faith, you must not get any ideas about this Mormon fellow. He may seem all right, but they have a terrible practice you don't know about. I have tried to shield you from such things."

"What terrible practice, Papa?" Her brow was wrinkled.

"Something called polygamy. They take more than one wife. Just a way to satisfy their lust, I believe. The good book says that a man and wife shall be "one flesh", not three or four, but one."

Faith sat tatting and frowning, listening to her father.

"So don't you go getting any ideas about him."

She didn't answer. Her father leaned his ample body nearer to her chair. "You haven't got any romantic ideas have you?" She shook her head, intimidated. "He hasn't, has he?" She still didn't answer or look at him.

"Cause if he has, I'll teach him a thing or two. Why, I'd break that young man like a matchstick," portly Reverend Henderson bragged. He warmed to his subject, and continued vehemently, "Just let him come courting around my little girl, I'd show him what's what. He's no match for a real man. Jacob Whitley could lay him out cold. In fact, I would, before I'd see him touch my little girl. Damned Mormon!"

At that, Faith's lifelong teaching of absolute obedience failed her. She threw her lace onto the floor and leaped out of the chair, fists clinched and turned on her father. "He's not a damned Mormon! He's a fine person. He's been nothing but nice and pleasant to you, as well as me. You have no right to talk that way. John Patrick O'Neill is twice the man Jacob Whitley is. And yes, I do have some ideas about him. If you really want to know what I think, I'll tell you. I think I love him."

She finished a bit breathlessly. The Reverend was clearly shocked. "Faith, I've never heard you speak like that. After all I've done for you! I can't believe what I'm hearing. Love? You love this Mormon fellow? You don't know what love is, girl. And look what he's done to you, inciting you to be disrespectful to your father."

"I know what love is, Papa. For the first time in my life, I know what love is! It's wonderful. It's being happy and laughing, being sweet and gentle, being thoughtful and kind. It's caring more about the other than about yourself."

"Faith!" Now the Reverend was truly alarmed. "When has he had a chance to 'teach' you all that kind of thing? Have you been seeing him alone?"

She backed down a little. "Well, a few times."

Her father was on his feet now. "How many is a few times? What do you do when you're alone?" He grabbed her shoulders, a terrible thought rising. "Has he compromised you?"

She stared at him in astonishment. "Compromised me? Why no, Johnny has never treated me with anything but respect. He wants to marry me!"

"Oh, my heavens! Oh, my good heavens!" He fell back into his chair. "A Mormon in our midst. I can't believe it! My little girl in love with a Mormon, and he wants to marry her. How many other wives does he have, the libertine? Oh Faith, Faith. I'm going to faint. Everything is going dark. Oh my good daughter, don't leave me. Faith, where is your hand? Please don't leave me. What would I do without my Faith, my own little girl? Where are you, daughter, where are you? Everything is getting dark."

The good Reverend was truly in an emotional state, and enjoying it fully. Faith knelt beside him, gravely concerned. She had never seen her father like this before, helpless and calling her name. She was concerned. Suppose she had caused a stroke or something terrible.

"Please calm down, Papa. I'm here. I'm right here. Now don't worry. Don't take on like this. I'm not going anywhere. Here, here's my hand. Papa, are you all right? You're not having a stroke, are you?"

"I don't know, girl. I truly don't know. My mind is all awhirl. Just the thought of you and that Mormon set me all awhirl. Promise me you won't do anything foolish like marrying him." He grasped her hand more firmly and passed his other hand over his eyes in anguish. "Promise me, Faith. I have raised you the best a father ever could. It would break my poor old heart. Promise me."

Faith looked off, down the road toward Johnny's place. "I . . . I guess so, Papa." And a tear rolled down her cheek.

Johnny had his house finished by August—at least, the outside structure and all the clay work. He had started on the inside, dividing up the space into rooms, picturing the bedrooms and the kitchen with Faith in it. He even plowed up land for a flower bed outside the front of the house. He was anxious to show it to Faith but had not seen her for two weeks. Three times he had gone riding over to Henderson's, but he had the distinct feeling he was unwelcome. The Reverend never had time for him, and Faith wouldn't come out onto the porch. The third time he asked her father if Faith was ill, and Henderson quickly answered no, that she had work to do.

He had also begun to feel a standoffish attitude with folks in the town— everyone, that is, except Hector. The two had become fast friends, and Johnny confided all his glowing plans for life with Faith with his friend. Hector had no woman. "Too big and homely for anyone to want me," he would say to Johnny. John frequently told him, "Out west, a man like you, big and strong, would be a prize, a real prize." And Hector would grin.

It was early August. One day Johnny had ridden into town and spent an hour or so talking to Hector in the blacksmith barn. The day had started fair and sunny, but by late afternoon, a wind blew up and the sky clouded over with deep, dark clouds. They began to hear rolls of thunder and an occasional crack of lightning. John stood up, saying he'd better get back out to his place, and make sure the lightning hadn't started any fires.

John rode a very nervous horse out of town. His mare was skittish, and he had to work to keep her on the road. The sky was entirely black now, and they picked their way along, with thunder goading them on. Suddenly, a loud crack startled John, and he ducked his head. He heard a strange whistling sound, as something went by close to his ear.

"Is that a shot?" he wondered. "No, just lightning, I think. It must just be my imagination."

Then came a second crack, and his horse reared and bolted. John hit the ground rolling. He rolled into the underbrush under the trees, and waited for a long time. This time he knew it had been a rifle. No one passed by. Still, he waited and watched. At last, he roused himself and made off through the trees to the Henderson place. He knew it was beyond that stand of trees, much closer than his place, and it had been two weeks since he had seen Faith. Besides, if someone was shooting at him he dared not go back to his own cabin.

He ran through the meadow and leaped up onto her porch, soaking wet and dripping.

Pounding on the door, he called out, "Henderson, open up." There was no answer, though he could see a dim light inside. He grew concerned. "Henderson, it's me, O'Neill. Open up, it's a terrible storm!"

For a moment Faith struggled with the promise to her father and her desire to be with Johnny. Lightning flashed and Johnny instinctively ducked, pounding on the door. Faith threw it open. He ducked inside and grabbed at the door, flapping in the wind. Pushing it closed and bolting it, he turned to Faith, "Where's your father?"

"He's gone," she whispered. "I'm here all alone."

John could see her fear at once. Her eyes were wide, and her face was absolutely white. She was holding herself with a tight rein. He stepped closer to her and took her head in his hands. Tipping her ashen face up to his, he spoke gently, "It's all right, darling. The storm won't hurt you. I'll stay with you. Don't be afraid."

She began to shake and sob, and threw her arms around him, not caring if he was wet. "Come, sweet, you sit down here by the fireplace." Johnny set her gently in a chair. "I'll get a fire going and get dried off."

"I already tried and the wind puts it out."

"Never mind. I've built fires in worse places than this. Be still now, honey, it's all right, really it is."

"I hate storms like this, Johnny. They scare me. When I was a little girl, lightning hit my uncle's house and blew it into a million pieces. My uncle was killed. It was terrible, terrible!"

He kissed her to stop the hysterical words. He kissed her again and again, every time she started to speak. Finally she put her head on his shoulder, and he held her as she calmed down. After a few minutes he knelt by the fireplace, where she had tried to start a fire, and rearranged the wood and the kindling. He took the candle, lit the kindling that he soaked in pitch, and soon he had a crackling fire. Steam went up off him as he knelt before the fire and held Faith's hand.

"Now tell me why I haven't seen you in two weeks. I'd begun to think you were avoiding me."

Her pretty face was a picture of woe. "I have been. Oh John, what can I do? A while ago I told Papa about us. He dragged it out of me. I didn't want to tell him, but he said some awful things about you, and I couldn't stand it. So I told him, and it nearly killed him. I think he had a stroke. He fell back into the chair and everything went dark around him, and he begged me not to leave him. After all, he really has done a lot for me, and he needs me so badly. He made me promise not to marry you."

Johnny was listening, picturing the Reverend working his daughter around. He raised his eyebrow. "So here you are, Miss Faith, with two promises. One promise to marry, and one not to marry. Whatever will you do?"

She didn't answer, just looked down at their clasped hands. "If you don't marry me, I'll have a stroke, too," Johnny said solemnly. "In fact, I might very well die. I tell you, everything is terribly dark without you.

I'm lonely as a polecat. I can't sleep. I can't eat. What will I do without my Faith?"

"Oh stop, Johnny. You're just saying that. You always tease me."

"No, my sweet thing. I'm not teasing a bit. Look at me, and see if you think I'm teasing." She lifted her eyes to his. Johnny was not smiling now. His blue eyes were not twinkling. They were filled with shiny liquid, and he raised her fingers up to his lips. "I love you, Faith. You are the only girl I have ever loved. You must be mine or I really couldn't stand it."

She threw herself into his arms, and they sat cuddled up on the fireplace hearth. "I love you, too, Johnny. So much, so very much. What can I do? Papa will never give his consent, and he's the only one around who can marry us."

"That's right," Johnny mused. "I hadn't ever thought about that. Kind of sticky."

"One thing he said that I don't know how to understand. He said Mormons have a thing called polygamy, and they have more than one wife, sometimes three or four. Help me understand that, Johnny."

He saw the anxious concern in her eyes. He searched her face and found it unbearably beautiful. "Faith, darling. I don't want anyone but you. In the Old Testament, God's chosen people frequently had more than one wife. Many societies still do. Maybe there are just more righteous women than men. Maybe they just needed to increase their posterity more rapidly. Solomon had many wives. David had many wives, and the only one he was not justified in was Bathsheba. Even Abraham had several wives, and of course, Jacob did. I don't really understand it myself, because there is only one woman I have ever found that I love and want to marry, and that is you."

"Do you really think the Lord justifies having more than one wife?"

"Well, he didn't seem to be against it. Abraham was one of the greatest prophets that ever lived. Surely, if God had disapproved, He would not have had him as a prophet."

"I suppose not." She was looking into the fire. "Maybe you'll want to have a lot of wives, too, some day. Mormon men do, Papa says."

Johnny caressed her long golden hair, lying in soft waves down her back. "There'll never be anyone as beautiful as you, and I'll never do anything to hurt you."

She turned to him and kissed him passionately. Her body was pressed against his. He could feel her softness and her desire. "I love you, Johnny," she whispered against his cheek. "I want so much to be your wife. Let's run away. We could go to another county and find a preacher. We could leave tonight, while Papa is gone."

Johnny took her arms from around his neck and moved away. He stood up, to try and think more clearly. Maybe Henderson never would let him marry her. Maybe she was right, and they would have to run away if they wanted to marry.

Johnny was standing in front of the window. Suddenly he heard something heavy on the porch, and threw himself on the floor at the far end of the room. At the same moment, a hard object began to pound on the door. Faith screamed.

"Faith Henderson," a deep, rough voice called out. "You got that Mormon fella in there?"

She didn't answer. She looked wildly from the door to John Patrick on the floor. Her hand was clamped over her mouth to keep from screaming again.

"Faith, it's me, Bill Wheelwright. Send that feller out here. We don't want no damned Mormons around this place."

Johnny crawled over to her. "I'm going out the back door."

"How do you know there isn't someone at that door, too?" she whispered anxiously.

"I don't. But I can't stay in here."

"I'll try to draw them to the front door."

"Damn it, Faith, open this door," the gruff voice called out again. "Where's your Pa? He wouldn't help no Mormon."

"Bill," she answered fearfully. "How many men have you got, or is it just you? He's kind of mean. I don't think you could take him alone." Johnny smiled at her inventiveness. He edged toward the back door.

"Hell, I could take him alone. But I got two more besides. Now send him out here."

"I'll try," she called out. In a loud voice, she said to the empty chair, "I think you'd better give yourself up. They aim to take you anyway."

"Well?" Wheelwright hollered.

"I think he's coming out, but he says he wants to fight all three of you in a fair fight, and if he loses, you can ride him out of town on a rail."

"If he loses! There ain't no question about it. I got Jim and Hank right here beside me. They're raring for a fight. Send that miserable no good out here."

"All right. Oh, Bill, he says, have you got a gun?" Johnny was long gone. He had slipped out of the back door, watching for glints of guns in the lightning flashes. He went straight back, past the clothesline, stopping under the trees and watching. As he left the trees and ran back through the garden, Johnny could still hear Bill hollering to Faith. After the garden, he knew it was too far for them to see him.

"Shore I've got a gun, and by damn, I'm about to use it, too. If he don't come out here, I'm gonna shoot the lock off this door, and you'd better get yourself out of the way, Missy."

"No, don't do that," Faith called in alarm. "He's coming. He swears he is, only promise not to shoot him."

Bill Wheelwright grinned and glanced at his two buddies. "I promise. Has he got a gun?"

"I'll ask him." They heard her voice saying "Do you have a gun, Mr. O'Neill?"

"No," she called back. "He is unarmed and coming out. Now don't shoot, Mr. Wheelwright."

Carefully, she opened the door and stepped back quickly. The three men burst past her into the room, The fire was blazing away, the chair was empty in front of the fire. They glanced wildly around the room. No one was there but the preacher's daughter.

Bill grabbed her shoulders angrily. "Blast your hide, you let him get away. What was you doing with him in here anyway, your pa gone and all?"

"Don't you grab me like that, Mr. Wheelwright. My pa will want to know what you are doing handling me like that. Mr. O'Neill sought refuge from the storm and only stayed long enough to build me this fire. Then he left."

"I don't believe it. You two search the other rooms. She's got a kitchen where he could be hiding and . . . " he smiled wickedly, "and two bedrooms."

Faith slapped him. "You're a vicious man, Bill. I never suspected it of you. You've known me all my life, and now you accuse me of something evil. My father will certainly be surprised at you."

Wheelwright looked down, and the smile left, but determination remained. "Sorry, Miss Faith. But I happen to know that scoundrel came in here. I heard your voices from the porch. And maybe you don't know Mormons like I do. They are scum. We rid ourselves of them a few years back. I never thought one would show his face around here. O'Neill's got a lot of nerve."

"That's not all. He's got a lot of good in him, too. You ask anyone in town that knows him. John is a decent, God-fearing man. Even Papa says so. He has a few ideas that are different from ours, but he's a very nice, Christian man."

Bill shook his head. "Mormons ain't Christians. They belong to a cult of the devil, gittin' our young folks to leave their ma's and pa's, following after that Joseph Smith fella. You know it has to be of the devil for him to get such a hold on their minds that they'd leave home and family for him. Don't you go a fooling around with that Mormon, Missy Faith, or you'll be sorry, sure enough."

Of course, the other two men couldn't find John, and the party of would-be lynchers went back out on the porch.

"Miss Faith, you best not let that fella around here anymore, especially with your Pa gone. Folks'll talk." With that parting advice, Bill and his friends left.

John Patrick watched them ride off, and waited for a good hour before going back to Faith. She let him in the back door and collapsed in his arms.

"I thought you did a great job. Who would have thought you could lie so well!" His eyes smiled at her. "Maybe I am corrupting you, Miss Faith."

"You're scaring me, mostly," she replied, her face against his wet shirt. Suddenly she pulled back. "Look at you, still wet through and through. You're going to catch pneumonia. Come over here by this fire and get warm."

She brought him a blanket and went out of the room, while he stripped off his wet clothes, placed them by the fire to dry, and wrapped himself in her blanket. It was late into the night before they stopped talking, and she dozed off to sleep in the chair. He watched the fire, keeping it stoked. When his clothes were dry, he put them back on, and settled down on the hearth again. He watched her sleeping. Her hair was in disarray around her shoulders, and her face looked tired.

Maybe I am wishing a hard life on my little one, John thought. Maybe I shouldn't ask her to share all my hardships. And they could get worse. I had hoped the hard feelings over Mormons were gone, but I was wrong.

All at once he had a strange vision of Faith walking beside a wagon, with nothing but hard baked ground and a huge, empty sky all around. Then it faded, and she was simply in the rocking chair, breathing softly. A kind of pity swelled inside him. A woman commits such an act of courage in giving herself to a man. "Poor little Faith," he whispered. "What are you getting into?"

Reverend Henderson drove up in the buggy before breakfast. He stopped short when he came into the house and found Johnny with Faith.

"Now, sir, I know this looks strange, but I swear to you I've only had your daughter's interest at heart. The storm was terrible last night, and she was awfully frightened of the lightning. Then, too, there were some wild fellows wandering around in the storm. They banged on her door and scared her to death. I just couldn't let her stay here alone." Johnny said it all in one breath.

Henderson looked from one to the other, and then back again. "Do you mean you stayed here all night with my daughter?" he breathed heavily.

"Well, sir, I had to, actually. She was simply not safe. I'm surprised, really, that you, such a doting father, would go off and leave her like that. She's so helpless, you know."

"Don't you call me a neglectful father, young man. Why, I've raised that girl since she was a baby. I've taken care of her all these years. I sure don't need you to help me or chastise me."

Johnny shook his head. "All I can say is, the circumstances speak for themselves. Where were you last night when poor Faith was frightened out of her wits by the lightning? Where were you when she was crying from fright of those three rough men?"

"Three rough men!" Henderson bellowed. "Faith, who were they? Did they hurt you? Tell me!"

She glanced at Johnny. "No, no they didn't hurt me. It was just Bill Wheelwright and two friends. They . . . they stopped by. They were looking for someone."

"That's right, sir," Johnny quickly reassured the anxious father. "But coming out of the storm like that and pounding on her door, they sure gave her a good fright. Then, too, she didn't have a proper fire. Couldn't get one started. I had to start up a roaring fire just to keep her from pneumonia."

A thought began to settle in Henderson's mind. "Did Wheelwright find him," and he pointed to John, "here with you, Faith?"

"Yes, Papa," she answered demurely.

"I'm afraid so, Reverend. Course they were concerned about her reputation, and probably rightly so. I didn't think of it then, but they cautioned us not to mention we had spent the night together. They were afraid people would talk."

"People! What about them. Three of them, and Billl the biggest blabbermouth of all. Why, Faith, your name will be blackened worse than that chimney."

The Reverend sat down heavily and put his head in his hands.

Johnny grinned at Faith over her father's head.

"Places I come from, they often put the two innocent young people through a shotgun wedding. I'm awful glad they don't believe in that out here. I mean, Faith and me, we're not even sure we would want to get married."

Her eyes widened, and she started to open her mouth. John motioned to her to be quiet.

Reverend Henderson moaned, "I can just hear the gossip now. It will ruin her. It will ruin me. Where were you when they came in?" he asked Johnny.

"Oh, me, I was wrapped up in a blanket most of the evening, drying my wet clothes by the fire."

"Oh heavens, oh my good heavens," Henderson moaned even more deeply. "It's worse than I thought. They saw that, no telling what they must have thought." Then the thought struck him. He stood up and grabbed Johnny's shirt.

"Did you touch my daughter, you, you Mormon, you?"

"Why no, sir. Of course not. I wouldn't do that. Then I might have to marry her, and I'm not aiming to get tied up with a preacher's daughter, of all things."

"What's wrong with a preacher's daughter? A good God-fearing woman would do any man good. What's wrong with my Faith?"

"Oh, not a thing. Only I'd hate to have to marry her."

"Oh you would, huh. Well, you should have thought of that before you came here and spent the night with her, and set the whole town talking. They'll probably run her into the ground, and I'll never have another good tent meeting in this entire county. You've as good as ruined her."

"I wouldn't say that. I've been completely a gentleman."

"Has he, Faith?"

"Certainly, Papa . . . all except for once," Faith said sweetly.

"Once!" he exploded. "What happened once?"

"It was unfortunate, but he was kissing me when Bill came in."

"Oh, no. Oh, no. Faith, my child, don't tell me that."

"I'm sorry, Papa. He promised he wouldn't say anything to anybody."

"Well, Faith, I'd better be getting on," John Patrick started for the door.

Henderson roared, "You get yourself back here, young man. You can't ruin a young girl's reputation and her father's position in the community, and just walk away. See here, you've got to marry Faith to save her face in this town."

"Now, hold on. I didn't think you were the kind to give shotgun weddings, and force two young people against their will." Johnny was enjoying this immensely, and Faith was desperately trying to hide her smile.

"I don't care what you didn't think. I will not be ruined in this community. I have built my whole life around preaching in this here county, and I'm not going to let a young whippersnapper like you destroy everything I've built. By heaven, you will marry my daughter, or you'll feel my buckshot."

And so it was that John got the desire of his life, and Faith Henderson, the preacher's daughter, was married off by her own father at a shotgun wedding. Johnny's little cabin was bare, except for one straw bed and a potbellied stove, a table and one chair. He had wanted it to be just right for Faith, but she didn't see anything except him anyway. He carried her through the doorway and set her down in the chair. There was still a dampness in the air from the storm that had just passed over, and Johnny built a small fire while Faith sat quietly watching.

Now that they were alone, at last, as man and wife, they were both silent and a little self-conscious.

"Are you tired, Faith?" he asked her.

"No. Johnny, stop messing with the fire and come to me."

Slowly, he stood up, wiping his hands on his pants. She had taken the braid out of her hair and shaken it around her shoulders. A little breeze from the doorway stirred it. He knelt down beside her.

"I don't want to scare you, Faith. Do you want to wait for awhile, I mean, til we get used to one another."

She reached out a small white hand and touched his cheek, lovingly. "I'm used to you right now. You won't scare me. I love you too much."

He pulled her down off the chair beside him on the blanket over the straw that he called a bed. And he was gentle. And he was infinitely tender. It was the first time for them both, and it was as natural as breathing, and as beautiful as a sunrise. Many years later, Faith would tell her children that marriage to John Patrick O'Neill was the dawn of her real life, and he would smile . . . as always.

CHAPTER 7

Shannon had sired half a dozen foals since Charlotte had first broken and trained him. Most of those she had also trained. She had nearly talked Jack into letting her train all the very young animals. She did so well with them, Jack realized it was good business sense to let her do it. The animals she trained were well-behaved for anyone who treated them gently. The only common trait they had was a very sensitive mouth that resented being sawed by an insensitive person. As long as a rider was light on the reins, Charlotte's proteges were eager and responsive. Her animals sold for twenty-five to fifty dollars higher than any others on the ranch. The buyers liked their calm, gentle spirits. Then, too, being sired by Shannon, their lines were purebred, and they were fine looking animals.

Jack drew the line at Charlotte racing Shannon. She wanted to ride him herself, but Jack said that she was not to make a public spectacle of herself. So Jimmy John raced him for awhile, until the boy turned into a man and weighed too much. After that, Emery's youngest son, Ben, began racing him. In a few years, the horse became unbeatable, both in Iowa and the surrounding states. Men from Chicago and St. Louis began using Shannon as a stud for their mares, and the horse was worth almost as much to Jack as he was to Charlotte. But not quite. Shannon was still Charlotte's best friend. There was a bond between them forged by lonely hours of solitary rides, grazing and dreaming by meadow streams, poking through the remains of old farmhouses, eating wild apples together. Unnumbered tears had dropped on the sleek, shiny, chestnut neck of the stallion. Charlotte liked to say that Shannon was a real gentleman. His eyes were soft and ever watchful, never wild or shifty. Unlike many stallions, he would stand perfectly still for a rider to mount him, and would respond to the slightest pressure from the knees. Charlotte took the finest care of him, brushing his coat constantly, cleaning his hooves and wiping the outside wall of the hoof with turpentine to keep the enamel strong and hard.

Charlotte and Jack fought about everything else, but about Shannon they found neutral ground. Indeed, it often seemed the only things about which they could civilly communicate were the horse and the ranch. When they would argue, Jack frequently seemed to succumb to a wild darkness that obliterated reason or care; then everyone stayed out of his way. Sometimes, after a storm between them, Charlotte and Jack would meet out in the stables a little later, and their anger and bitterness would fade a little while brushing and talking about the animals. Still, he would not take her anywhere with him. He always said she had so offended his townfriends that she would not be received. Finally, in September, he took her to Carthage with him on business. She made the most of the occasion, dressing in her finest frocks, laughing and teasing with Jack all the way, and becoming the center of attention amongst his associates in town.

On their way home the following day, Jack was in a black mood. After several attempts to make conversation, Charlotte said, "What's wrong? Didn't you have a good time?"

"Hell, no!" he growled. "I went to do business . . . which I couldn't do for watching my wife flirt with my business associates."

"Oh, horsefeathers! I didn't do anything of the kind. Can't I talk to anyone without you thinking I am flirting? Besides I seem to imagine you got all your 'business' accomplished."

He stared straight ahead, fuming. After a minute, Charlotte put her hand on his arm. "What did you want me to do, Jack?"

"I wanted you to sit there and look pretty and shut up."

"Now you know that's not me! I never have been like that. You married me because I wasn't like that."

"Well, I didn't think you'd act like a common woman."

"Can't I talk to anyone but you?"

"No, by damn! Leastwise not the flirty way you were talking. It'll be a cold day before I let you embarrass me again."

She stared at him, not believing that he could be so jealous. But she soon was forced to believe it because all the comings and goings around the ranch soon excluded her again. When buyers came, she was not introduced nor even told. When she passed the ranch hands, they looked the other way.

The only people she had to talk to other than her children, Matt and Ruby, were Johnny and Faith. At least twice a week, she would ride to their cabin and sit with them. It was hard for her. They were so happy, and so obviously in love. She saw between them that which she would never have with Jack and wondered how they could be so lucky. They, too, compared. When Charlotte would go, Faith would say to Johnny, "Can't we help her? Can't we help her get away from him? I've only met him twice, but he scared me both times."

"I'm afraid she made her own bed, and she is too stubborn to give it up. I tried to get her to go with me out to Utah, and she wouldn't. She said she'd have to leave her children, that he'd kill her if she tried to take them."

In the spring Charlotte became pregnant again. It was secret for a couple of months. Finally she told Lily.

"I'm about three months gone, Lily."

"I 'spected it for a good while now. I seen you was blooming in the face and neck. I says to Emery just a week ago Satidy, 'I reckon Mrs. Boughtman is in the family way again'. Well, well, that's fine news. Be a Christmas baby, will it?"

"About then, I expect. I don't know if I can wait that long. It's a good thing we haven't got any say over it, or I'd probably crack it's shell like a baby chick and spoil the whole thing."

"How's Mr. Boughtman feel about it?"

"Well, you're the first one I've told actually. I'm waiting for the right time to spring it on him. I imagine he'll begin to notice pretty soon, anyway."

Lily looked at her closely, narrowing her eyes thoughtfully. "You'd better tell him. Never know when a man would need to know that kind of thing."

Charlotte knew she was thinking about eight years ago when Jack had kicked her, and she had almost lost Matthew because of it. She made a mental note to take Lily's advice, but she didn't get the chance.

She dressed up one fine April day, took Ruby, and went riding into Montrose. Charlotte had decided it was time that she made her peace with the town. Shannon was hitched to the carriage, and people turned their heads to stare at the big stallion meekly pulling a wealthy woman and her little girl. She tied him to the post outside the general store, and she and Ruby went in.

"Good afternoon, Mr. Mickleson. I'd like to see your yard goods and toilet water, please."

"Certainly, Ma'am. Right this way. I have some more things in the back I'll bring out when you're done here. Pardon me, Ma'am, but I don't think I recall your name. Clumsy of me, a lovely lady like . . ."

She turned her most charming smile on him, her gray eyes flecked with sea-green. "I'm Mrs. Jack Boughtman, and it isn't at all clumsy of you. It's been a long time since I was in. She patted the cloth before her. "I intend to be a regular customer from now on, though."

He mumbled some words of welcome, and stumbled over his feet in getting to the back, where his wife was piecing a quilt. "It's Boughtman's wife," he whispered to his wife, a big, red-faced woman with pin-pricked fingers. Both of them crowded around the curtained doorway, trying to see without being seen. Charlotte was not fooled. She had peeked through enough curtains herself to be aware when she was being watched. Neither was she concerned. Let them watch. That was why she had come, to be seen and to meet the people of the little town. After looking over his goods for the better part of a half-hour, Ruby was tugging on Charlotte's dress, so she bought several things and asked to have them delivered to her ranch. She said warm good-byes to the stupefied storeowner and his wife, and followed Ruby outside for a social walk down the street—the only street of the town.

Ruby had been to town many times, and now she served her mother as a willing and self-important tour guide. She pointed out the livery stable, the boarding house where she and Matthew and Jack had sometimes had lunch or late breakfast. They stopped in and met Miss Ann who ran the place. She was a nervous, flutter-puff of a woman, constantly moving from one knick-knack table to another. Still, she was very nice, and it was so good to be in the company of other women again that Charlotte forgave her the eternal fussing and throat clearing.

By the time they went back out on the street again, the word had spread that Mrs. Boughtman was in town, and many of the ladies were either peering out of curtains or standing in their doorways. There was one nine-year-old boy, hobbling as fast as he could move with his lame leg, toward the town's only saloon, at the opposite end of the street. Charlotte was not particularly inclined to look in on the saloon. She knew it would not be the proper thing for a lady to do. She was about to turn back to her carriage when she heard the laughter. It was loud and deep, unmistakably the gruff voice of Jack Boughtman.

"Let's go find Daddy, shall we?" she asked brightly of Ruby, and the little girl was eager to agree.

As they crossed the street in front of the saloon, she saw something through the open door that caused her to stop. Turning to her daughter, she said, "Ruby, I'm worried about Shannon standing for so long at the other end of the street. Why don't you go and stay with him a minute."

"I want to get Daddy."

"Do as I ask, please, or you'll not come to town again."

Ruby looked longingly toward where her father was intermittently laughing, but she stamped her little foot and went reluctantly to Shannon. She passed the lame boy about halfway down the street.

Charlotte walked slowly up to the doorway of the saloon. Inside, she could see Jack. His back was to her. Beside him was a girl. She wore a simple, thin, white blouse with a very low, rounded neckline. Her skirt was dark blue, and it fell softly about her hips. There were no petticoats to make it stand out, and, obviously, no corset underneath either. As Charlotte stood looking in, Jack's hand made it's way beneath the thin, wavy dark hair. He pulled the girl's head back and kissed her roughly, then tasted her neck and shoulders as well.

The little lame boy had reached the doorway, too, and stood looking up at Charlotte fearfully and sorrowfully. They both stood there a moment, she watching the scene within, and he watching her. Carefully, he edged his way through the doorway and hobbled up to Jack. Timidly he touched his shirt.

"What d'ya want, Danny, can't ya see we're busy?"

Danny couldn't get the words out he wanted to say. Instead, he simply pointed. Jack's gaze followed his outstretched finger and found Charlotte on the other end. The girl, Danny's sister, turned also and paled when she saw Charlotte.

For all her anger, Charlotte never dreamed, at first, that there was really anything between Jack and the girl. He took his loving so regularly from her that she had never given any thought to his having another woman.

"Is this your 'business'?" she asked coolly.

"It is today." Jack replied defensively.

She scanned the girl contemptuously. "It must be a very boring day if this is all you can find to do."

The girl spoke up in a throaty voice, slightly slurred by whiskey. "We spend a lot of boring days together. At least he seems to be bored when he gets here. After awhile, things get real exciting."

"Shut up!" Jack said. "What's up, Charly? You're all fixed up like you're going to a wedding. What are you doing in town?"

"I came to shop and socialize and, perhaps, ride home in the company of my husband, who was supposed to be here on business." As she continued, her eyes began to spark and her old brogue crept back. "I see, however, ye are too occupied with yere business acquaintances to spend the afternoon with yere wife and daughter. I hope ye enjoy 'his' company, Mr. Boughtman, for ye might as well stay the night and the morrow and forever, if ye please. I do na' want to see ye in me home again."

She started out the door, but Jack had her by the arm, with one great leap. "Don't get your Irish up, Charly. Me and Constance are old friends, and don't go telling me whether I can come to my own house or not."

By this time, her voice was low and barely more than a hiss. "Get yere hands off me." Her fingernails dug into his hand, and he yelled and back-handed her. Charlotte had taken many such blows; but, in her anger, she was not prepared for it, and she fell against a table, wrenching her ankle as she stumbled. The table went down with her as she glanced off the side. Ruby came dashing in just then to find her Daddy. They all stood like statues for a minute, Jack unprepared for the chain of reactions. He really hadn't meant to hit her so hard. Whiskey had dulled his sense of proportion. Danny was terrified at the sight of this fine lady falling, with a big blue bruise spreading itself across her cheekbone. Ruby looked at her mother sprawled out on the floor like a big, fancy ladybug, all puffed up with skirts and parasol, and, inexplicably, she was ashamed. She turned and ran outside.

Jack tried to help Charlotte up. She wouldn't let him touch her. The boy, Danny, helped her into her carriage, and she drove away from Jack while he was still calling to her. Where to go? Not home, she would not go to Jack's house! Tears were streaking her cheeks as she urged Shannon ever faster. She came to a fork in the road, and she knew that, just through the trees to the left, she would come to Johnny's house. Heedless of the effect it would have on him, she knew she had to find refuge somewhere. He was her only strength.

Charlotte's carriage came racing in and pulled up short. Faith saw it through the open doorway, and putting down the spoon she was stirring dinner with, she hurried out. Charlotte was ashen, her clothes in disarray, and her face disfigured by an ugly black and blue bruise. Faith reached up in compassion for her, and Charlotte came tumbling down. Faith almost fell with her weight, but steadied them both and helped Charlotte inside, laying her gently down on the bed.

"Oh, Charlotte, Johnny's not here," Faith said. "He's been gone all day to Hector's, helping him with some things in the blacksmith shop. I don't expect him 'til dinner."

Charlotte looked up at her sister-in-law, her eyes dull, her face a mask. "It's all right, I'll wait. I'm good at waiting."

Faith sat patting her hand, and putting cool cloths on her face, until it seemed Charlotte had gone to sleep. The younger girl stood over her and shook her head. Then she turned back to the stew she was preparing.

After looking all over town, Jack found Ruby sitting on a tree stump down by the boarding house. Her face was streaked with tears, but she dried them when she saw her father coming. He came up and stood in front of her, holding out his hand. She glared up at him defiantly, until he said, "Come on, Missy. No reason to pout. Your mother got just what she deserved. She was rude and embarrassed me in front of my friends." Finally the little girl accepted that answer, and gave her hand reluctantly to her father. He put her behind him on the horse and headed home.

But Charlotte was not there when he rode in. He could see that instantly, for the carriage and Shannon were still gone. He put Ruby down and turned the horse around.

Emery appeared in the barn door. "You seen Charlotte?" Jack called out.

"No sir, Mr. Boughtman. Ain't come home yet."

The answer came swiftly to Jack. He wheeled his horse and headed out the gate. He had never been to John O'Neill's place, but he knew where it was, and he knew Charlotte went there often. It was her only diversion lately. Easily he found the fork in the road, and galloped through the trees into the clearing.

Charlotte heard the thunder of horse hooves and raised up, hoping for Johnny, but somehow knowing it was not. Faith didn't have time to get outside. Jack came barging in, his head barely clearing the door frame.

"You got my wife, girl?"

Faith backed away from him, too frightened to speak. He looked down at her white face and neatly coiled hair. She was too timid to even interest him.

"Where is she?"

Still Faith didn't answer. Scowling darkly, Jack reached out and grabbed her wrist in a flash. "You can't hide her, you know. She's coming home with me. Now, do I have to tear your sweet little home apart?"

Charlotte knew that Faith was no match for her husband. She had struggled to her feet in the bedroom and now moved slowly into the doorway. Jack turned to her and dropped Faith's arm. In two strides, he was at her side.

"Charly, you look sick. Come on, I'm taking you home."

"Not going with you." She shook her head weakly but full of determination.

"The hell you're not. You think your brother and his little missus will protect you from me? There's nothing they can do to keep a man from his wife. Now, come on, Charly, I'm gonna take care of you."

Charlotte started to protest but Jack swooped her up in his arms and turned for the door. There stood Faith. In her hand, she had a fire poker.

"Put her down, Mr. Boughtman. You've hurt her enough. She came to me hurt and sick, and I'll take care of her."

Jack stared in disbelief. Then he laughed short and harsh. "You! I don't believe it. You little scrawny, white-faced calf. Get out of my way, and put down that poker before I brain you with it."

"Jack," Charlotte cried at him.

He looked into her face, wincing inside as he saw the bruise he had caused. "What," he said gruffly?

"Leave her alone. She's just trying to help me. Faith, please, please, just let us go. I'll be all right. It wasn't his fault," she lied. "It was mine. I fell and hit my cheek on a table. Please, Faith, and don't tell Johnny."

But, of course, she did. As soon as he came home, Faith ran out into the yard and caught his reins.

"What's wrong," John asked, surprised at her fervor.

"Johnny, Charlotte was here. She was sick or something, and there was a big bruise on her face. Terrible. It looked awful, and she could hardly walk! I laid her down on the bed and tried to take care of her, but her husband came. He grabbed her up and took her home. I know she didn't want to go. I even tried to stop him, but he just shoved me aside. I'm scared of him, Johnny. He is capable of anything. He could kill her some day."

That was all he needed to hear. He wheeled his horse around and started out at a dead run. What usually took a half-hour took him only fifteen minutes, and he went racing in through Boughtman's gate.

Before he could even dismount, Jack was on the porch.

"No need to get off the horse, O'Neill. Just head right back home. Charlotte is my wife, not yours. And right now, she is upstairs in bed, sick. She don't want to see you or nobody else. The doc is with her."

Johnny could see another carriage over by the stables. "If you've hurt her, Boughtman, so help me I'll kill you."

"Oh you will, huh! That'd be a tall order."

"Charlotte doesn't deserve to be treated the way you treat her. I know all about you and your temper, and your other women. The law won't protect you forever. You can't treat a woman like a dog. If I find out you've hurt her, I'll find a way to legally take her away from you."

Jack laughed and waved him off. "Get off my place, O'Neill. And don't come back. You can't scare me with your talk. You go on home and take care of your little woman, and I'll take care of mine. And, by the way, don't expect any more visits from your sister. She's gonna be in bed for awhile."

"Damn you, Boughtman, what's wrong with Charlotte?"

"Nothing too much. She just lost a baby, that's all. Doc'll take care of it."

John sat for another minute, staring up at Charlotte's window where he could see the outline of a doctor bending over. Finally, there was nothing else he could do, so he turned his horse around and reluctantly rode away.

Charlotte had hemorrhaged and lost the baby. That night in her bitterness and anger, she taunted Jack until he blew up and admitted that, yes, the girl in the bar was his mistress and had been for several years. She lay against the pillow listening to him with dulled senses, bitterly remembering how confidently she had told Johnny she would never share her husband with anyone. She said nothing, while he raved on until his black rage was spent. Through it all, Charlotte remained as a stone, uncaring. Her baby was gone—so was her pride. The one ultimate reason she had for rejecting her family's religion and life was a mockery. Everything seemed to be stripped from her, even her will to live.

But Jack would not let her die. He labored over her long after the doctor had gone. His rage spent, he sat all night watching her face, bathing it when perspiration stood out on her skin and, all the while, damning himself in his mind; all the while, cursing himself for loving her when she didn't love him.

After that, it was dark in the place where Charlotte lived. Not the black of Jack's angry world, but a kind of shadowy twilight where she moved in almost a dreamlike trance. Memories became her friends and ghosts her confidants. In her mind she had many long talks with Johnny and Annie, and many remorseful tears were shed on her father's red, beefy neck. Margaret became the voice of reason and understanding, and now Charlotte listened as she had never done as a sixteen-year-old.

What vivid scenes came washing over her as she sat alone in the rainbow-prismed parlor of her home—visions of Margaret washing her long brown hair over the wash tub and braiding it like a crown on the top of her head—visions of Patrick bursting open the door of their little home hollering, "Where's my queen and the little royal shirt-tails?" And she saw Annie and John Patrick, rushing at their father to be the first to have their ears pulled and noses tweaked. Back in the shadows, she saw herself watching, with wistful eyes, the generous love that fed their family.

Where was it now . . . all that love and laughter she had thought would always be there? Where was the sunshine she remembered gilding Annie's sweet baby face? Vaguely she knew it had to be somewhere, but she couldn't find it. Sometimes she would ask "Is the sun shining? Where's the sun?"

Lily would tell her, "It's just outside. Why don't we take a walk in the sunshine, Mrs. Boughtman?" But, cruelly, when she ventured outside to see, there were only shadows and twilight.

After that day in Montrose, she rarely went out, even to see Johnny. Not only had Jack forbidden it, but the sweetness of Johnny's life with Faith was almost too painful to watch. He came to see her sometimes, always when Jack was gone.

She asked John once how he always knew when her husband was away. "Oh, I have my spies," he smiled, not admitting that Emery was his best source. Charlotte did not press him. She was desperately afraid Jack would hurt him someday because of her, and she warned Johnny constantly to be careful and to take care of Faith. She loved Faith, in her own way. It would have been hard not to, for the younger girl was so full of compassion for her and so full of adoring love for her brother. So, it was partly her fear for them and partly the shadows of her mind that kept her home.

Jack was completely confused as to what had happened to Charlotte. He hardly believed that the knowledge of his unfaithfulness could cause this change in her. He could rant and rave as he wished, and she only watched him silently. When his frustration mounted in her new retreat, and he would shout at her, "Answer me, damn it," she would quietly say, "I have nothing to say to you." It was a new Charlotte, one he could neither understand nor reach. His rages of anger and frustration became empty lashings of a storm on a deserted shore, and that angered him even more. It was the old Charlotte he loved, one who would resist his imperious will, one who would stand and give as good as she got. Where had she gone? And why had she gone? He had thought she would soon return to storming and yelling and berating him as she had always done, but this time there was no storm. In fact, she hardly seemed to distinguish when he was there and when he was not.

Her craft was broken, beached on a shore he could not fathom, and he was alone. And Jack hated to be alone.

Dinner was over. Faith had cleared away the dishes and swept the hearth. Night had begun to come earlier now that September was here, and she and Johnny liked to sit in front of their fire, making plans for their coming family. Faith was four months pregnant and her belly was just slightly rounding, a fact which John loved to tease her about. When they had been sure of the good news, they drove over to the Reverend's house in their buckboard one Sunday to tell him he was to become a grandfather.

Grandfather Henderson was resigned to the marriage by this time, even though it caused him some inconvenience. Faith still went back twice a week to cook and clean for him, and he was frequently a guest at their house for dinner. He and Johnny had some wonderful discussions about the spirit of man, modern prophets, and 'faith versus works'. Johnny loved to say that, since you can't be saved by only one or the other, he was glad to finally have Faith so that he could now concentrate on a little work.

They sat before the fire, Faith leaning back against her husband's shoulder and Johnny stroking her long hair, feeling beneath it her blossoming body. He nuzzled her cheeks and neck and shoulders.

"Has there ever been a wife so loved?" she asked wonderingly.

"I doubt it," he said, coming up for a breath. "Although 'Faith' has been sought after by all mankind."

"Silly!"

"Sorry."

"I know. You can't help it," she said in mock exasperation.

"Your father shouldn't have given you such a funny name."

"He didn't, my mother did."

"Well, see. It's her fault, not mine."

"What will we name our baby?" she asked.

"What are you going to have, a colt or a filly?"

"Let's hope for a boy. Boys are more helpful to their fathers around the farm."

"Yes, but girls are more helpful to their mothers around the house."

"True."

"And maybe she would look just like her mother. I'd like that." He pulled back her hair, and kissed her behind the ear.

There was a rustling outside and Johnny sat up. Then there was the sound of boots, and he could catch a glimpse of a flickering light. John grew alarmed. He called out. "Hector, is that you?"

"Hell no!" came the reply.

John pushed his wife down on the floor and, grabbing his rifle, hid beside the window. He glanced out but saw nothing in the dark, nothing except more flickering lights. Torches, he thought, and he was right.

"O'Neill, we're here to tell you to leave Montrose. We don't want your kind around."

"What kind is that?" Johnny hollered back. "A hard working farmer! You got something against that?"

"Ain't talking about farming. Talking about Mormons."

"Who are you?" Johnny yelled.

"You know me. I tried to get you to leave a year ago. I thought you'd take the hint. Now I ain't hinting, I'm telling. We all come to tell you. It ain't just Bill Wheelwright, it's the whole town. Go on out west with your Mormon friends. Go preach your religion to the Indians."

All at once Faith was beside him at the window. He motioned for her to get down, but she was not daunted. Faith was a very gentle woman, but with righteous indignation she could be as bold as any man. She stepped out into the doorway and called out.

"Bill, you come on up here so's we can see your face. Yes, you and any other man that calls himself a man. Now aren't you ashamed, trying to frighten a man and his wife off their land. Why, I've grown up with most of you as neighbors. Come on out of those shadows and talk to us like decent folks."

Slowly and reluctantly, several men came out of the bushes, holding rifles and torches. They gathered around the porch, surly and a little sheepish.

Johnny was standing beside Faith, his arm around her, his rifle on the floor beside them.

"Miss Faith . . . I mean Mrs. O'Neill, ma'am, we don't mean to harm you . . . " Jim Stanton started to explain.

"No, but you'd kill my husband and burn my house? That doesn't make sense, Mr. Stanton."

"He's a Mormon," a man in the back shouted out.

"And what's that, Mr. Beel? I've lived with him for a year, and he's the kindest man I've ever known. He's no devil. If you'd get to know him, you couldn't help but like him."

"Shh, Faith," John silenced her gently. "You men have come here with a grievance against me, now let's hear it out. What's your grievance?"

No one spoke. There were about ten men, and they shifted about uncomfortably. They were all men Johnny had met and shaken hands with in his year in Montrose. They never said anything unpleasant to him on a one to one basis, but now, in this crowd, they had grown a little ugly.

"I'm not converting anybody. I'm not preaching to anybody. I'm not bothering you or your families, and I obviously don't have horns. What are you unhappy about?"

"You ain't a Christian," one voice called out.

"I'm sorry to disappoint you, sir, but I am. I love and believe in Jesus Christ and in his Father."

"You got some kind of a golden bible that you believe in over the real one," another man spoke up.

"Again, you'll be disappointed. I love the Bible with all my heart, and the golden bible you talk of is simply a written record of some people who lived on this continent many thousands of years ago. I love it, too, but it sure doesn't take the place of the 'Good Book' as you call it."

There was silence. Bill Wheelwright glanced around at the men's faces. They had begun to soften. "You can't let him talk us out of this. You know if we let one Mormon in, the rest will come in droves, just like they did over in Nauvoo, and we'll be shoved off our land. The Bible says Satan will come as a ravening wolf in sheep's clothing." He pointed at Johnny. "He's a sheep now, but what about when your sons and daughters are running off to Utah to join up with them Mormons. What about when your own good wife leaves you to follow this man's crazy religion." He was playing to their fears. And it worked.

"Leave town, O'Neill," a man shouted angrily at him.

Johnny shoved Faith back inside. As he did so, he felt a rough hand grab his shoulder and swing him around. Then a heavy fist smashed his jaw, and he saw Wheelwright's face grinning above him. Faith screamed out and tried to get to Johnny, but two of the men pushed her back inside and closed the door on her. The men had become enraged, feeding on old hatreds, old fears and the excitement of the moment. They dragged him off the porch, pinning his arms, and dragging him by the hair toward one of their horses. Johnny was a big man— almost six foot three—and strong, but he was no match for the eight men clawing at him, wrapping his ankles with rope, and kicking him in the ribs. Bill jumped into his saddle and spurred his horse, all the while holding the rope that tied Johnny's ankles. The horse bolted forward, and John was being dragged behind, bumping over the hard ground, sticks and grass tearing at his back. To keep his head from being smashed on the ground, he curled forward so his back took the punishment instead. Somewhere in the distance, he could hear Faith screaming, and cursed the man who would scare her so. He kept working at the rope with his hands, trying to shut out the stabbing pain in his back and ribs. They seemed to be going round and round the house. He could see the torches whirling by him and men's faces staring down at him. Fighting off unconsciousness, he had finally freed one of his ankles when the horse came to a sudden stop, and Bill tossed the rope down on him.

"Better git, Mormon. Next time, we'll shoot you and burn your house. Won't be no reprieve." Wheelwright led the pack out. The other men rode home, but Wheelwright rode straight to Jack Boughtman's and reported their success to him.

John lay curled up in pain and heard the hoof beats fade down the road. Then Faith was kneeling beside him, cradling his head in her arms. Her tears dropped down on his cheeks.

"Oh, Johnny, my love, my darling. How could they do this to you? You don't deserve it. How could they, how could they?"

"Get the rope off," he croaked out.

She worked the rope loose from the other foot, then helped him drag his bleeding body back into the cabin. He stretched out on his stomach before the fire, and she bathed his back until late into the night, then wrapped him in bandages. She was crying, but she was also angrier than she had ever been in her life, angry at injustice and stupid prejudice. She never wanted to go back into Montrose again. And she never did.

They sold out, for much more money than Johnny had paid—as he had once predicted—and packed up their few belongings. They stopped at the Reverend's to say good-bye. He begged Faith to stay, but to no avail. Johnny went into town one last time to say good-bye to Hector. He found his friend packing a buckboard pulled by two horses.

"Going with you," Hector announced in his deep voice. "If you'll have me, that is. You said a big man like me would be a prize out west. Well, I might as well find out." John grabbed his friend's arm and was too choked up to say anything except, "Let me help you."

Last of all, Johnny stopped at Charlotte's to give her one final chance to go with him. Her eyes were dull, her manner listless. He hardly knew her.

"Char," he said, "You can come with us. We'll be in St. Louis before Jack will even miss you. You said yourself he is often gone for days. He'd never find you in all that crowd of people."

Charlotte gazed at him hopelessly, "What for, Johnny. It's no use. You don't understand, I'm tired of living. There's nothing to live for anymore. Let him do whatever he wants with me."

Johnny felt so helpless and exasperated. He had never seen her like this. Nothing he could say would move her. Tears stung his eyelids. He remembered the Charlotte he had grown up with, and he saw her now, broken and hopeless, looking years older than she should have.

"Good-bye, Johnny. Say good-bye to Faith. Take care, won't you."

Tears streamed down his cheeks as he bent to kiss her, but she was calm, resigned and had no tears to offer him.

"I'll write," he said. "And I'll come back. This is not the end, sis. I'll be back someday, I just feel it inside."

Charlotte smiled gently up at him, "Oh Johnny, I have loved you so dearly. Be happy. Oh, be happy."

CHAPTER 8

The one real pleasure Jack had in life was in his children. When Matthew was fifteen years old, he was almost as tall as his father, and his hair was a dark brown; but his eyes were Charlotte's—gray with little flecks of gold. His temperament was more like Margaret's, calm, watchful, and of few words. He had seen much in his young life. Jack took him almost everywhere he went. Matthew was there when Jack made business deals and frequently listened to Jack's philosophy of how to get the most money for horses you'd like to get rid of anyway. And Matt heard many of the royal battles that came from his parents' room. Often, afterwards, Jack would put his arm around his young son and tell him that women were hard creatures to understand or to tolerate. As the boy became ten and eleven, Jack began to warn him not to marry a woman who had an eye for the men. Matthew didn't know what he meant at first, but he was gradually led to believe that his mother caused the fights in their home by her obstinate ways and her flirtations with men. She was very careful, and Matthew never actually saw the men she sneaked out to see, but obviously, that was what drove his father wild. Matthew wished with all his heart that his mother would defend herself against the accusations that his father made, but she never did, so he concluded Jack must be right. In his boyish heart he grew sorrowful, believing that if his mother didn't love his father, she must not love him and Ruby either; for weren't they all a part of each other?

And yet, it didn't make sense. Most often his mother seemed distant and dreamy to him. At times she hardly seemed to know him, and other times he would see her watching him with sadness. Then she would come to him when he was sitting at the dining table and put her hand on his shoulder. She would rub his back or his arm and say, "You're growing up, Matt. I miss you." He didn't know what she meant. How could she miss him? He was right there at home with her all the time. But as he grew older, he no longer responded to her caresses with a hug of his own. Her attention was so sporadic, he grew suspicious of her affection and believed she was only pretending to love them.

Ruby, at twelve, cared for no one but herself and her tall, handsome father. She didn't understand how her mother could take such little pains to be attractive to him now. Jack loved to laugh and go riding, but her mother never went with him anymore. That was all right with Ruby. She would take her place. Ruby hung around the ranch, learning about the animals so she could be with her Dad. To be forced to stay inside the house and do woman's work was punishment to her. Her mother sometimes came out of her dream world to insist that Ruby learn to do needlepoint and sewing. She would make her daughter follow poor Sophie around, learning how to clean house and keep it tidy. But Ruby would so upset Sophie with her careless ways and sloppy work that finally Sophie would tell her to run off and play.

Why should she do housework? Her father never noticed what the house looked like, and he certainly never cared whether she did needlepoint, for goodness sake. It was much more fun to work with him, grooming or training a new animal. In the summer and fall, Ruby could scarcely be kept indoors. During the cold winter months, Charlotte once imagined she would spend some time taming her daughter, and she treated Ruby as she had Shannon, so many years ago. Tenderly, carefully, she held out her apples.

"Ruby, dear, after we do your lessons for the day, we could make you a lovely new dress."

Ruby would look at her with surprise. "Why? I don't need a new dress. I haven't any place to wear it."

Once Charlotte had invited her daughter into her world. "You know, when I was a young girl, I had only an old mare to ride, but we went all over the meadow, and I taught her to jump. My sister, Annie, and I used to build dams in the meadow stream."

Ruby had listened perfunctorily for a moment, then jumped up saying, "That's nice mother, but I've got to go out to the stable and help Matt."

Charlotte was more pained by her daughter's rejection than she had been by Jack's violent rages. She didn't know that Ruby had decided that day in the saloon never to be a fancy dressed, prim and proper lady.

She had also decided that her mother was despised by her father, and, therefore, she, Ruby, would not have him think that she was at all like Charlotte. When she looked in the mirror, she saw her own looks reflecting Jack's dark handsomeness. Her hair was jet black, and glistened down to her waist. Her eyes were so dark as to be black, and the lashes rimming them were long and dark. Her mouth was wide and tended to be thin and precise. Only her skin was like Charlotte's, creamy white.

There were nights when Charlotte dreamed that she lay beside Patrick in a cold, deep grave out on the Western Plains. Overhead she could hear howling of the wolves and the thundering of the buffalo hooves. And she was afraid. But her father would speak and say, "Charlotte, me girl. Do na' be afraid. I'll not let anything hurt ye." And her fears would vanish. The dark coolness of the grave would seem a blessed balm to her tormented soul. She would call out to her father, "Papa, can you hear me? I love you, Papa, I always loved you." Then she would turn over and he would be gone. She was alone in the grave, and the howlings became louder and louder, until she woke with a start, staring into the shadows of her bedroom.

She often talked to Johnny in her mind, and heard him say to her, "I'll come back someday. This is not the end. If you ever need me, just tell Him, up there, and it'll get around to me. You know I have never let you down." Those words came back time after time to her, and how she wished she could get to Johnny. It would be all so simple, and so good, but she couldn't bring herself to pray. How could she go to the God she had rejected in her foolishness and pride? How could she go limping back to Him, beaten by her own stubbornness and rebelliousness, and expect He would now take pity on her? Why should He? She had rejected Him, and it seemed He had likewise rejected her. Never had she apologized for herself the many times she was punished by Patrick for her rebelliousness. And now, in her hopeless, barren world, she could not go to God with her apologies. She would accept her punishment. And so she was frozen in her misery, bound by her self-conviction, immobilized by her pride.

One night Charlotte sat in the high-backed, satin chair that Jack had just had shipped in from Chicago. All the house was dark. Emery and Lily were retired to their rooms, and Sophie was asleep in her quarters. Charlotte had accepted a pitcher of ice water from Sophie just before the servant retired, hours before.

The darkness seemed to taunt her with twisted, evil faces. The friendly shadows of her day-time world frequently gave way at night to other faces, grotesque and terrifying. She often sat like this in the dark, watching those faces, holding them off simply by the will that was within her. Jack taunted her for lack of backbone. He never knew the raw courage it took for her to face the night, to lie awake with those forms, gnarled, gnome-like, leering at her. For as long as she could keep her eyes open, she lay staring at them, facing them down like you might a pack of snarling dogs. They were all around her, leaping, reaching, grinning their hideous grins at her. She sat in her chair as still as possible, for as long as possible. So long as she sat perfectly still and faced them, they did not touch her. She feared to move, because then she wouldn't know exactly where they were. Perhaps they might touch her with their awful, bony fingers.

At last, weariness overcame her. She had to get to bed. So, unsteadily, she pushed herself up from the chair, exerting all the will she possessed to force those images to move from her pathway. She made it to the parlor doorway. Then, at the foot of the staircase, she stumbled and fell. She thought she saw one of the little creatures beneath her foot, and she screamed in terror at the thought of pinning him beneath her. In falling, she ripped her dressing gown, and her hair tumbled over one bared shoulder. It was at that moment that the door opened on Jack and Matthew.

They had been in town for a special dinner at the Masonic Lodge. They were both dressed in their ruffled, silk shirts, looking every inch the gentlemen. Charlotte looked up, tears running down her face from fear, her hand over her mouth, praying that she would not be sick.

"Good heavens, woman, do you have to subject your son to this kind of sight?"

Matthew stood rooted to the spot, his eyes like saucers. For a second, they grew glassy with little-boy tears, then he, too, became contemptuous, imitating his father. Charlotte saw her son looking down at her.

"Matthew, I'm sorry. I didn't mean to . . . "

He turned and walked outside. He had been struggling with his desire to love his mother and his father's constant humiliation of her, but now he understood that she was mentally off. Tonight was proof—his final proof. She was weak and unbalanced, and he was ashamed.

He sat outside in the porch swing until he heard Lily's voice coaxing Charlotte to come upstairs. His father came out a minute later. When all was quiet in the house, Jack said, "You go on up to bed now, son. I think I'll sleep in the bunkhouse."

Matthew looked at him in pity. "How long has she been like this, Pa?"

Jack looked at him seriously for a minute. "Well, she kept it hid pretty good for awhile from me and you kids, but she's getting old now. That's only one of the things that has driven me to Connie. Your mother has always been slightly deranged. Some times are worse than others, and she's always leaned on other men. She's broke my heart with it many times. Be careful of women. Don't ever let them get the best of you. If you let 'em, they'll take advantage of your good nature. You're too gentle, Matt. A woman will do you anyway she can. Get you a good, loving woman who'll listen to you, and then keep a tight reign on her. Just like a mare; a mare ain't really yours 'til she's broke good."

Days went by, nameless days—not dark, not light—gray days, lost forever in the shadows of Charlotte's mind. She did only what she had to do, receiving no joy from living. Life was a burden, and days turned into years. Her eyes lost their sparkle, and she never whistled, no matter how sunny the day. Winter was the same as summer to her, and even Shannon could not bring a smile to her lips. It was as though she ceased to care and only moved like a slave through the work of living. Finally, one August when Matt was sixteen, Charlotte roused herself from the shadows. She had lived there long enough, and the unquenchable spirit within was tired of fear and burdens and pitiful shadows. A happy memory and a tenuous hope was working in her mind.

"Lily, when I was a girl, I was made to do needlepoint until I could scarcely look another needle in the eye. I hate sewing. I hate housework. I hate being a woman. I should have been a man so I could ride, and work, and do business, and get something fun and worthwhile out of life. Well, I can't work, and I can't carry on the business myself, but I sure can ride!"

She bathed and fixed up her hair and dressed in her riding outfit that had hung in the closet for three years. She ventured out of the house into the bright sunlight of the June day. It hurt her eyes, but at least she saw no shadows. She saw no leering shapes, no creatures before her path. She hurried to the stables. There was Shannon. He whinnied to her and nuzzled her, when she held out the apple she had brought. He hadn't been curried in many weeks, and his flanks were flabby from lack of exercise. Well, she would remedy that. She spent an hour brushing his warm, silky coat; then she cinched up the saddle she always used on him.

"Come on, old friend. We are going riding, we two. It's a sunny day, and I'm sick of the house. Let's go." She headed him toward the river, and they crossed on the ferry. Once on the Illinois side, they went cantering leisurely down the deserted roads that led to Nauvoo. On the other side of the ghost town was farmland, grown over with weeds, and bare except for a few farmhouses that had sprung up there in the last few years. They stopped when they came to the lane that had once led to the O'Neill place, now barely recognizable in the undergrowth. Very slowly she turned Shannon down the lane, and he plodded along until he came to what once was a clearing around the house. Now the grass grew almost three feet tall, and all that was left of her childhood was an old chimney and hearth with bricks broken out in places. She sat lost in thought, seeing visions of the people who had walked here. How could the joyous promise of life at seventeen have turned to dread? With heavy heart, she headed the horse toward the meadow where she had spent so many delicious hours. It was too late in the year for berries, but the flowers were blooming. She dismounted by the stream and let Shannon drink and wander as far as he would, chomping lazily on the grass. Charlotte lay down in the long, sweet grass. Oh, the smell of the meadow—flowers and cool grass and the rich earth beneath her back. Here was peace of mind and serenity.

After a time, it seemed she could breathe again, deeply, as she once had done. The blue of the sky and the sunshine seemed to penetrate the fog of hopelessness that had shadowed her days. That same indomitable spirit that had allowed her to fight Jack's power for so long seemed to gain strength from the sunshine and the fresh air.

Memories came sweetly now, memories of happy times, and with them the courage to face her mistakes. "I've paid so dearly for my pride, Papa," she said out loud to the blue, sun-filled sky. "My willful ways, as you called them, have led me to the valley of the shadow. I don't like it there, Papa. I was so afraid. Why do we have to suffer so to learn? Isn't there any other way? I guess there isn't, for me. It takes a lot to 'learn me'."

Charlotte's mind became filled with all the warnings she had been given about Jack. She saw again, as if it were only days, instead of years, the eyes of Joseph Smith looking into her soul. She heard his voice predicting she would some day be happy to be a tenth wife to a good man. And she heard his fateful words, "Life is a challenge and a choice, and beware the wrong choice." Tears streamed unrestrained down her cheeks. How blind she had been. How headstrong and smug to think she alone was right in the face of all evidence. How could she ever have thought Jack would confine his appetite to just one woman?

The whole foundation for her rejection of her family's religion had crumbled, and in retrospect seemed pretty flimsy in the first place. It was almost laughable. She had never really had her husband all to herself. She had always shared him with his super-human egotism and selfish will.

She lay for a long time, numb with pain, as she saw herself again at seventeen, rebelling and hurling accusations at her father, condemning him for not trusting her, refusing to believe him when he warned her against Jack. But no, she would have her way! No one would be her master! No one would tell her how she must live her life—not her father, not Joseph Smith, not even God! Sorrow came rushing in like the wind reaping the field, and she fell into a kind of stupor, no longer suffering, but simply suspended in time. It was long, long hours before a realization came stealing over her soul, that she also had the power to overcome the awful existence to which she had doomed herself by her willfulness. At first, it was simply a wispy thought, but it kept returning and growing stronger. Finally, late in the day, when the sun had disappeared, and the sky was laced with ribbons of gold, pink, and gray, she began to arouse herself. The thought came so powerfully to her mind—you can be happy if you want to, or you can wallow in your misery. She sat up in the grass and stared into the sky. "I want to be happy," she said aloud. "I want to be happy and, by heaven, I will be happy, despite Jack, despite anything."

Shannon was grazing nearby and whinnied when he heard her voice. He came over to where she was curled up in the grass, and she rubbed his nose. "No, we're not going home, not yet. I don't want to go back into that house again. Let's just sleep here tonight, underneath the stars. You won't mind, will you, boy?" So they did.

Charlotte slept that night by the bank of the stream, curled up snug and content, with her head and shoulders just underneath Shannon's chin. Several times during the night, her hand went out in her sleep to touch the sleek powerful neck of her friend, lying beside her in the grass. She saw no imps that night. She had no nightmares. She felt safer outside with Shannon beside her than in her own bedroom with Jack.

When the morning came, she drank long and deep from the cool, clear waters of the stream where she once had bathed. The dam she and Annie had made was long since gone, and the pool much reduced in size. So, as the sun crept up in the morning sky, she began once more the enjoyable task of building another dam. It took little thought, but kept her busy and gave her a purpose for that one day. Late in the afternoon, with her riding pants wet clear up the thigh, and her back stiff and sore from bending and carrying, she sat back to survey her day's work. It was a fine job, and the finished dam gave her great satisfaction. It made a small waterfall and several feet of rapids. Her hands were red and wrinkled, and her feet were bone-cold and aching.

At last, she became restless. For one thing, she was hungry. Shannon had been munching on the tender grass and drinking from the stream, but her stomach had been empty for two days, except for the water she had drunk. The restlessness extended to a growing desire to prove to herself that Jack could no longer hurt her with his unfaithfulness, his violence, his constant battery of contempt. She determined that day, she would never allow him to drive her that far again.

It was late afternoon when she rode into the ranch. On the large, oak dining table at the house, there lay an invitation to the Governor's Ball. It was being held at Quincy, some forty miles south, where he had his private home. Governor Hamilton of Illinois was inviting Mr. and Mrs. Jack Boughtman to be in attendance at the Governor's Banquet and Ball in three weeks. She pondered over it for awhile, then went looking for Jack.

He was in the barn with the vet. One of his mares was going lame. Charlotte had changed her clothes and repinned her hair. Jack didn't acknowledge her presence until the doctor was finished. The two men started to walk out together, but Charlotte said, "Jack, I'd like to ask you about something."

He glanced her way impatiently and reluctantly said good-bye to the vet.

"Just where have you been, Missy? Matt tells me you haven't been home since yesterday morning."

"It doesn't matter. Obviously I haven't made my escape from you," she replied sarcastically. "I noticed the invitation to the Governor's Ball. Are we going?"

He looked at her contemptuously, "I am accepting the invitation. It's good business. Unfortunately, my wife is indisposed."

"Actually, I'm feeling fine," she replied. "And I think a Ball would be very pleasant. I'd like a new dress for the occasion."

He looked at her for a minute, then turned on his heel and walked away. "No," he said, over his shoulder. "I can't risk having you make a fool of yourself and me in front of all those important people. Go back to your dream world, and stop trying to be something you're not."

"Then you won't take me?"

"No. I wouldn't dream of taking you. It's too fancy for you, and too important to me to mess it up."

He left two weeks later on a business trip to Springfield. He was not coming back until after the Governor's Ball. Charlotte suspected he was going to Springfield mostly to outfit himself in expensive, new clothing for the fancy occasion. One morning while she was exercising Shannon, the thought occurred to her that Jack would be mighty surprised if she simply showed up at that Ball. The more she thought about it, the more intriguing the idea became. She would simply come like Cinderella to the Ball, and let the devil take the hindmost. A marvelously delicious feeling tickled her, and she turned the horse abruptly toward home. When she let the reins out, the horse ate up the miles home in a few minutes.

She pulled Shannon up in front of the house. Calling for Ruby, she ran into her home. In a moment, the girl came to the top of the stairs, and Charlotte called out, "Go walk Shannon out, would you, honey. I've just given him a long run."

Ruby pouted, "Why can't you do it? You worked him up."

Charlotte smiled at her and answered, "Because you're going to do it. I have some other things to do. Right now, please, he's panting out there."

Ruby grumbled and mumbled, and dragged her feet down the stairs. While her daughter was walking Shannon, Charlotte was on her knees in front of the old cedar chest that her wedding presents had been stored in. She now used it primarily to store linen and quilts. Still, in the bottom, folded as neatly as the day she had first seen it, was the bolt of blue, silk cloth Patrick had offered as his "sorry" token. She had sworn, in her anger, she would never use it. Now she took it out and unfolded it. It was smooth and luxurious, covering her arms and lap with easy elegance. Where Patrick had gotten that cloth, she did not know. It was, undoubtedly, the finest piece of cloth she had ever seen. Her heart began to swell with memories of her father, and she realized how much he had loved her. The whippings had faded away these last few years, and she remembered only the times he had humbled himself, expressing his love and sorrow to her. She sat for a time wrapped up in the silken cloth, caressing it longingly as though it were her father's face. Finally, she folded it up and took it with her as she went looking for Lily.

"We'll make a dress that will knock the Governor's eyes out." she said. "I'll show those good Illinois people that Mrs. Charlotte Boughtman is not some crazy person. And if Jack doesn't like it, he can go jump."

Lily sucked in her breath and glanced around to make sure no one had heard. She thoroughly disapproved of Charlotte's plan. A woman should not go against her husband. The man was the head of the house as Christ was head of the Church. Wasn't that what the Good Book said? Charlotte was going to be sorry. So she mumbled through the five days it took to make the dress. They had it done three days before the ball. It was magnificent! The sleeves were full and gathered into three puffy tiers right to her wrist, where a ruffle discreetly hid the upper part of her hand. The neckline was low enough to be alluring, but modest enough to lend dignity to her status as a married woman. The bodice fit her snugly, and the skirt fell over her petticoats in large, soft folds. She surveyed her reflection in the mirror with satisfaction.

It was a two day's journey, and since she couldn't go alone, she tried to persuade Emery and Lily to go with her. Emery was adamant. He would not go. Mr. Boughtman would be furious and he would not risk his employer's displeasure. Ben might drive her, but how would that look for her to go off across the country with a young, seventeen-year-old boy. So in the end, Charlotte persuaded Lily to accompany her and let Ben come as their driver.

They stayed in a rooming house in Kingsborough and arrived in Quincy by four in the afternoon the next day. The two rooming houses of the town were full of the elite society of Illinois coming for the Governor's Banquet and Ball. Charlotte sent Ben inquiring where they could find accommodations. The old maid, Heloise Pardish, had turned her old family home into a rooming house for the grand occasion. She had one room left with a double bed. Charlotte and Lily got that. Ben slept in the livery stable.

Charlotte was careful to wear the veil of her hat over her face, even in going from the carriage to the house. She knew none of the people around could possibly know who she was, but she knew, also, how rumors and idle remarks spread. So she disappeared into the room and set about freshening up. Her dress was unpacked and the wrinkles steamed out of it. She lay down for a little while with cool, wet cloths over her eyes. By five o'clock, the town began emptying a parade of silk and organdy, cigars and leather, color and chatter, and the noise of "importance." Like multicolored grains in an hourglass, the great folk trickled out of town into the Governor's mansion two miles away.

Charlotte hadn't the nerve to present herself unexpectedly at the banquet table. She calculated that the most auspicious time she might appear would be directly after the banquet, just as the dancing was beginning. So she took a small, plain meal in her room to avoid the curious questions of Miss Pardish. Already the woman had been following Lily about, trying to wheedle out bits of information from her. Lily was as discreet as she should have been. The most the old maid found out was that Charlotte was a very rich woman from Iowa. It was exasperatingly unsatisfying to her gossip-steeped soul.

At eight o'clock, Charlotte could bear the room and the wait no longer. Ben brought the carriage around. Then she and Lily slipped out right under the very nose of their hostess. Charlotte made Ben trot Shannon past the lane leading to the mansion. They could see the mass of lights from the white, manicured home, that was set off to advantage at the high point of the green slopes. She held her pocketwatch in her hand, glancing at it incessantly, trotting up the road and back again, until the hands crept agonizingly to the half-hour position. Sure enough, the orchestra was tuning up. When Shannon trotted up the long lane that circled in front of the mansion, the butler called a carriage boy to park the vehicle with the others, but Ben handed Charlotte out and resumed his seat. He and Lily drove down the lane, now fast darkening with the shadows from the overhanging trees.

In her red-gold hair, were half a dozen blossoms entwined and tucked into the curls. In the bosom of her dress was tucked a white gardenia. The royal blue of the dress gave her eyes almost a summer-sky appearance. Dressed in her beautiful blue gown, she moved like a queen.

The butler brought her into the ballroom and announced her. She stood quietly, poised by his side, secure in her appearance, gracious in her acceptance of the stares and whispers. The Governor came quickly forward.

"Mrs. Charlotte Boughtman," the butler said with dignity.

The orchestra paused just before beginning the first piece, and Governor Hamilton hurried to take her hand.

"Mrs. Boughtman, what a pleasure! Indeed, what an honor! We are so happy that you were able to overcome your illness. My dear, Mr. Boughtman will be most pleased, and we," he gestured to the increasing number of gentlemen who were edging over for introductions, "consider ourselves privileged, indeed."

Her voice was soft and inflective, "You honor me, I'm sure. Your invitation was simply too kind and enticing to let any illness keep me away." Her eyes were a constant fluid movement of eyelash and sparkle. "I only regret that my decision to come, and my recovery so sudden that I was not able to properly respond to your invitation. I do hope I haven't inconvenienced you."

"Oh, my dear lady, nothing could be further from the truth. You so brighten our event, that your presence is quite a treat and not at all inconvenient. Before these gentlemen fill your dance book, I wonder if I might have the honor of the first. But wait, I am forgetting myself. Mr. Boughtman will most certainly claim the first dance. Very well, then, the second, most certainly, must be mine."

She looked past his shoulder and appreciative eyes. Jack had been engaged in conversation on the other side of the room and had not noticed her entrance, but now he was crossing the room, all the while watching the cluster of men that surrounded her. She could tell by the set of his jaw and the hardening of his face that he was angry. No matter, she had known he would be. She now had to prevent him from forcing her outside. As the men stepped back to allow her husband access to her, she slipped her arm discreetly through the Governor's and felt him start with surprise.

Jack spoke pointedly, "Charlotte, my dear, I thought you were ill."

Before he could go on, she inserted her excuse, "Oh well, it was nothing serious, thank goodness, just a touch of the summer vapors I suppose. I knew how upset you were when you had to leave without me, so I determined that nothing should keep me from your side."

She turned her wide, blue-gray eyes on the Governor. "Governor Hamilton has been most gracious in receiving me. In fact," she said, half to Jack, half to the Governor, "he has already claimed the first dance, though, of course, he was courteous enough to allow it to you."

Jack was furious. He could feel the anger growing red hot. She was acting the flirtatious wench again. He would be disgraced in the eyes of these men whom he had sought out so diligently. He hardly saw the gracious picture she presented. He was too wrapped up in his jealousy and anger and the realization that she had rebelled against him. Still, he could hardly be impolite to the Governor.

"Governor Hamilton," he said, bowing slightly and stiffly, "You do me honor to so honor my wife. I'm sure I could not deny you the first dance when Mrs. Boughtman and I have had many such times together."

She smiled at him, looking straight into his eyes.

The Governor responded, "Why, thank you. Most generous. It's a privilege to begin the dancing with such a lovely woman." He led her out onto the floor where the dancing had not yet started. The guests were awaiting his signal that it was time to begin. He drew her out with a flourish, her skirt swirling softly, the slight flush of excitement glowing in her neck and face. They dipped and swirled to the waltz, twice around the ballroom, the guests all whispers.

Finally, the floor began filling with other couples. Jack's first partner was the blonde daughter of a business acquaintance from Chicago. He took her back to her seat before the dance was quite over and waited to see where Charlotte would alight. When the music stopped, the Governor was thronged by gentlemen asking to be introduced and to fill her dance card. Jack looked on, unwilling to be a part of the crowd of men to whom she tossed favors and smiles. It was humiliating.

"My dear, we must introduce you around. The ladies would love to meet you, too, I'm sure. Where's that husband of yours?" Governor Hamilton twisted around in both directions, scanning the crowd for Jack. "Oh, there he is." He raised his hand above the throng and beckoned for Jack to come over. The big man moved hesitantly forward, reluctant to jump into this situation without knowing what tack to take.

"Now gentlemen, you'll all get your chance to dance with the lovely Mrs. Boughtman. But right now, these two people need to be properly introduced to our circle." He put an arm through Jack's and one through Charlotte's. Working their way along the perimeter of the room, the Governor introduced the couple to all the wealthy and important people of the State.

Jack was forced to smile and reply graciously to the many compliments of his wife's beauty. Gradually, he began to glance at her often himself, and admitted silently that she was unusually lovely that night. In fact, it had been years since she had appeared so attractive and full of spirit. He was on the alert for any slip she might make. He need not have worried. The years had taken their toll on her emotionally, but the last three weeks, without the influence of Jack, had been a time of renewing and determining that life should not beat her. Tonight her old, high spirits had returned. She was at her best, nodding and smiling modestly to the ladies, accepting the compliments of the men with discreet pleasure. Finally the presentation was over. Charlotte's dance book was filled. Not one place remained open for her husband.

So it began, Charlotte dancing all night with a different partner each time, Jack cutting in as often as he properly dared. When he was not dancing with her, he was watching her. He tried dancing with other women. There were certain ladies that it was socially expedient to escort to the dance floor. He did his duty. Still, his eyes scanned the floor to follow his wife, and as his anger cooled down, his disappointment took over. When she danced with a stranger, she laughed and smiled, sharing amusing conversation. Whenever he cut in, the smile faded and their time together was strained and silent. He complimented her. He remarked on her gown, on her hair, on the smell of the gardenia about her. She would not meet his eyes.

"Charlotte, look at me, damn it," he finally ordered in exasperation.

"I'm dancing with you, isn't that enough?" she said coolly.

"No, I want to talk to you, not at you."

She smiled, unconcerned. "Well, I have nothing to say to you. I am here against your will, and let's not pretend it is different."

After a minute he admitted, "I'm glad you came. I know I wouldn't bring you, but I didn't think you'd conduct yourself so well."

She smiled for everyone to see, but her voice was ice. "Disappointed, my dear? Your superiority is hard to see now, isn't it?" For him, her eyes didn't sparkle, they glinted. "If you had ever let me, I could have made you proud of me long ago. Now if there is pride, it belongs to me, because none of this came of your effort. I did it all myself. I owe you no thanks, and no love, and believe me, you won't get either."

The challenge hung there between them as a small, dapper man tapped Jack on the shoulder. Jack held her eye as she moved off in the arms of the State Senator. He stood watching her dip and turn, the soft light of the ballroom bathing her hair and skin with soft golden lights. She had been a useless, unwanted possession when she had lost her nerve and her fight. Now, once again, she was a possession to be proud of.

The evening was almost over, and Charlotte was still riding the crest of popularity. With the excitement of her first real Ball, she was not even tired. The polite conversation, however, was becoming a little tedious and her thoughts began to wander. She wished that Annie could be with her now. She wondered what her mother would think of all this, her mother who had never known anything except work and smoky, log houses. John Patrick would be proud of her, and she smiled at the thought. She wished he were here to see her and to dance with her. The only man she loved, and he had to be her brother. The Senator was droning on about society in Washington when a new kind of loneliness came sweeping over her. Jack was beyond love; he was to be lived with and survived. If she was strong, she might avoid being consumed in his will. But love, where was she ever to find it? Would it ever be for her?

At that moment, she glanced away from the Senator. She heard him ask if he could fetch her some punch, and she said, yes. But she didn't see him go. She turned slightly to her right and stood entranced as she looked across the portico. There was a man resting easily against one of the pillars. He seemed to be standing in a pool of light that was streaming from the doorway. He appeared to be possessed of a quiet self-assurance. He was dark-haired, with eyes as deep as the ocean, and he was watching her. She was not self-conscious under his gaze. Rather, she felt herself growing in beauty and glowing with a pride and self-confidence. They stood across the room looking at one another. She wanted him to look at her, to see no one else, to be taken with her as she was with him. A stirring began in her bosom. She felt her heart beating loudly and fast. A mysterious silence had descended over the whole ballroom, and she heard no noise, no music, no voices. The world seemed to have come to a standstill. There was only the dark-haired stranger watching her and smiling, and she wanted only one thing, to be near him.

Charlotte started toward him. She moved in a golden aura and saw reflected on his face the same intense tide of joy that was now enveloping her. It was a kind face she saw, strong but gentle. He might have been a poet, or a saint. For Charlotte, the world had receded and every other concern with it. Nothing existed except this man and herself. Just because she had never known true love didn't mean she couldn't recognize it when it came. It was all quite different from her challenging standoff with Jack. It was as though she were seeing, again, a beloved friend whom she had lost, and whom she loved intensely. Recognition of a kindred spirit was instant. Tears of relief came flooding to her eyes— relief from her prison of loneliness and hurt. Yes, love could come to her, even her. In a flash, she saw herself again at seventeen, kneeling beside her bed with Annie asleep, praying for help. That same feeling of peace and all enveloping love was flooding over her now from this stranger.

He was smiling at her but neither of them spoke. Their eyes held each other tenderly, and the years melted away until she might have been a seventeen-year-old girl again, meeting her soul's one great love. She smiled back shyly into dark eyes, soft and kind and deep as the sea. Was she beautiful enough for him? She was anxious to be only perfect in his eyes.

Heedless of the people about her, heedless—even forgetful—of Jack, Charlotte would have gone straight to his arms, as though to a safe harbor. He was stretching out his hand to her, and he smiled a welcome. But just then, the Senator spoke at her elbow. "Your drink, Mrs. Boughtman." She glanced at him, startled, and almost breathlessly accepted the glass. When she turned back, the stranger was gone.

Anxiously she asked the Senator, "Who was that man here by the pillar a minute ago?" He looked closely at the pillar and glanced about briefly.

"What gentleman did you mean? I didn't notice anyone by the pillar. Mr. and Mrs. Hartly are seated over there. Is Mr. Hartly the one you mean?"

"No, no," she shook her head impatiently. "There was a dark-haired gentleman standing against that pillar. I thought for a moment I knew him. He might be an old friend."

"Is that why you moved? I almost didn't find you again," he said reproachfully.

"Oh gracious, I am sorry," she put a hand on his arm, remembering her social manners. "I must have forgotten myself. You see, I thought he was a childhood friend. I don't see him now, though." She turned the conversation back to Washington politics so she could think her own thoughts. She determined to ask the Governor about the man, but she never had the chance. Jack cut off her dancing before the Ball was over, insisting politely, concernedly, that her health might be damaged if she overdid herself and stayed longer. She stepped hard on his foot.

"I feel fine, really I do. I'm having such a good time, I'm sure it can't harm me."

"Governor Hamilton will agree with me, I'm sure," Jack persisted. "Mrs. Boughtman is a lovely little butterfly, and no one enjoys showing her off more than I, but it would be a shame if she should relapse."

The Governor was all concern. "Quite so, my dear. You have graced our little Ball, and we are appreciative. We wouldn't have you ill because of it for the world. We want you available to come again."

Both Jack and Charlotte smiled, pleased at his enthusiasm.

"Thank you, sir," Jack bowed slightly. "Now I'm afraid I must get my wife back. I can always tell those small signs of fatigue. She takes such little thought for her own care that I have to." He sighed slightly. "I suppose, however, that husbands are meant to do exactly that, care for and protect these lovely ladies."

Charlotte thought she would be genuinely sick at his outrageous lies. Jack took her arm, escorting her out, guiding her firmly under the elbow. The stable boy fetched his carriage.

The night air was slightly cool and refreshing after the ballroom. As Jack clucked to the horse and they began their ride back to town, he asked, "How did you get here? I hope you didn't make that long trip alone."

She came out of her reverie. "Why should you care? Lily and Ben came, too." Then quickly she added, "But I don't want you to be angry with them. I made them."

He watched her face from the corner of his eye. "You surprise me, Charly. Just when I think you've lost all your spunk, you bounce back. This is my girl. This is the red-headed little girl that went bathing in the meadow."

She was beginning to feel the tug of fatigue once the excitement was removed. "If you like me so well like this, why always try to break me?"

He was amused. "I don't know what you mean. No one breaks you except yourself. I'll never understand you. One time you can be lively and fun to be with, other times you let yourself go, and I give up on you."

She was incredulous. "Jack, do you really not see what you do to me? You treat me like one of your mares. It's the sting of the whip if I show too much spirit, and the spur if I resist. I'm tired of living like that! I'm tired of always having to be strong, of having to fight you in order to survive. I want a man who will love me."

In the darkness, with the clip-clop of the horses hooves, Jack spoke softly, "I love you, Charlotte."

She didn't reply. She couldn't. Her mind could not digest what he had said. That he was serious was evident, and she knew deep inside that he did love her in his own way. But what a terrible, twisted way!

"No, Jack, too much has passed between us, too much hurt, and I can't always be the bobcat you want for a companion. I want different things now than I did at seventeen."

"What do you want?"

"A gentle man, maybe a poet, maybe a saint, but someone who will let me love him with all the sweetness that I can summon. I want a quiet love, a comfortable love, where there is talking instead of yelling, kindness instead of slapping. I don't know, Jack. Maybe I even want God."

He laughed shortly, "Now you're becoming sentimental just like any other woman. God! Then I guess you sure as hell don't want me."

Quietly she answered, "You are the one who determines that."

He stopped the horse and turned to her. His voice rising with emotion, he said, "Well, I want you. I've always wanted you, and you are mine. Don't forget it. You're my wife and I'll have you whenever I want you."

Carefully she shook her head. "No. You'll never have me. You can beat me, you can force me, but you'll never have me, the real me, because you don't want that. All you want is an anvil to strike your fire on. I can be that if I have to, but inside there's another me, and that person you will never touch."

He dragged her roughly into his arms, and his mouth closed on hers. She struggled but he was too strong for her. He kissed her mouth, her neck, her shoulder, and whispered, "Be my girl, Charly. Be my girl again." But she didn't respond.

"Remember that day in the meadow?" he whispered to her. "That first day and how good it was between us? I want my wife back. I've missed you these last few years. You know that nobody else means anything to me. Why'd you let it get you down about that other girl? You're my only one, honey. Be my girl again."

At last, she answered his coaxing. "Loving you is like loving a thunderstorm. I never know when it's going to strike me."

He laughed in amusement and kissed her again. When he had finished, he said, "You liked it once. You will again."

But it was several months before she would allow him to love her. Even then, she wept when it was all over. She stared into the darkness for a long time and thought of a dark-haired, soft-eyed stranger who had seemed to embody all that she craved of gentleness, tenderness. Where had he gone? What would she have done if he hadn't left? Would she still be Jack's wife, or would the man be dead from Jack's violent jealousy? Now that she had glimpsed a different, tender kind of love, the absence of it was the sharpest pain she had ever known.

She conceived again after a few months, and Jack pampered her as he never had before. Just before Christmas she gave Jack a healthy, laughing, red-haired baby boy. She wanted to name him John Patrick, but, of course, Jack would have none of it. They quarreled over it and finally settled on John Paul. He was the object of his mother's devotion and his father's amusement. Jack and Charlotte found in him another area in which they had no contention. Jack objected only to Charlotte's spoiling the little boy too much.

Even Lily said she spoiled her children. She had never whipped one. No matter how saucy Ruby was, she never felt the switch as her mother had. Matt was never outrightly disrespectful. He simply ignored his mother. He was a boy of sixteen now and a young man of few words, especially where his mother was concerned. He had been convinced by his father that Charlotte was a fickle wife, and that most women were like her. With his constant criticism of her, Jack effectively deprived her of the respect and affection of her son. As he drew away from her, he lost his boyishness, and began to sleep in the bunkhouse with the ranch hands. That way he rarely had to see his mother.

If Charlotte had not had John Paul to occupy her thoughts, she would have been more hurt over Matt's withdrawal. As it was, she was soothed by Jack's assurance that it was normal for a boy at this age to gravitate to other men and away from their mothers. She sometimes went out and watched Matthew train the horses. Once or twice, she offered him some good advice, but he rarely took it.

But the role of model husband was hard on Jack. After all, it was merely a role he played for amusement. He won Charlotte over, or so he thought, and soon the efforts to be agreeable became too much of a burden. Gradually the quarreling began again, and by the time John Paul was two, the house was electric with arguments. Mostly they fought about Ruby.

Ruby was growing up much too fast. She was almost fourteen years old now, and her little girl's body was fast turning into a woman's. She was becoming a real nuisance around the ranch, teasing the hired men with flirtations. Charlotte spoke sharply to her about it.

Ruby told Jack about the scolding in a little girl's voice. "Mama's always mean to me," she cried petulantly.

Jack and Charlotte quarreled that night until midnight over their daughter. Ruby crept down the hall and sat in the shadows beside the doorway listening. Charlotte told Jack he had not seen his daughter bathing in a long time, and he had to take her word for it that Ruby was fast becoming a woman. Ruby grinned in the darkness.

Her father was still the most attractive man Ruby had ever known. She was not afraid of him as most people were. Sometimes she sensed that even her mother, for all her backbone, was afraid of his black temper. It wasn't that Ruby never felt the brunt of his anger. She had also received her share of slaps, and more than once was sent sprawling across the yard for her insolence. She was not afraid of her father, but she was shrewd. She learned how to spot the signs that a mild storm was turning into a black rage. She could see it in his eyes. They became almost glassy and far away, as though he had lost his senses. Then she either disappeared or agreed quickly in her most honey-combed voice.

As for her mother, Ruby still held her in contempt but never underestimated her as a rival. And for all Charlotte never hit her as her father sometimes did, she had a way of cold, steely resolve when she gave a command that was challenged. Charlotte could usually quell Ruby's insolence with a long, calculated stare. Before the calm, unbending direction of her mother, Ruby gave in. She hated the feeling that Charlotte could see clear through her and mentally hold her against her will.

Other than Ruby, Charlotte and Jack had serious quarrels only about business. Of the two, Charlotte had the better business sense. Jack knew how to get what he wanted and for a good price. He didn't know how to keep a good customer. He never treated others with deference to their taste or opinions. He forced his own will on everyone, and he was not above passing off a bad horse on an unsuspecting buyer. Expert horsemen came to him because they knew that with care, they could get an excellent horse. Novices came because they had heard he had good horse-flesh. Unfortunately, they often came away with lesser quality than they might have had at another place. Charlotte knew, from overheard conversations among the ranch hands, that the Boughtman name was not particularly respected in business circles, and she knew with an inner instinct that was not good business. It was shortsighted business to cheat people. And so they argued.

One week they had quarreled about the sale of a mare that was going lame to an unsuspecting sharecropper some miles away. Charlotte had been angry about it in the beginning. Then when the man came back within a few days demanding his money or a new horse, Jack had him thrown off the ranch. She had watched it all from the porch. Hurrying to the box where Jack kept his cash, she counted out the money for the horse and slipped out the back way, crawled under the fence and ran after the man. Jack was furious when he found out.

Ruby told him. She had followed her mother when she had spotted her crawling the fence. Jack accused Charlotte of everything from stupidity to adultery. He finally slapped her, when he could win the argument no other way, and stalked out of the house about midnight. There was no doubt in her mind where he was going. Connie was still in town and Charlotte knew it, but she put it out of her mind, preferring to ignore it as long as he kept quiet about it. Not that she would ever be reconciled to that reality. If she had truly loved him, she would have fought it with all the spirit in her, but it had become just a minor irritation in her stormy marriage. Her mouth quirked grimly when she remembered her idealistic opposition to polygamy in her youth. Now she lived with a sordid counterfeit of it.

The following week Jack stayed in town every night, coming to the ranch during the day to work the horses. Charlotte managed to be gone every day riding Shannon. She would take John Paul and seat him before her, and they would roam the meadows and streams all day, wading in the water, eating berries where they could find them, sometimes taking some of Sophie's biscuits and pork for lunch. The little boy adored his mother, and he loved his father with the same fervency. Jack was unfailingly kind to him, and he missed his father that week.

After the household was in bed at night, Charlotte would walk. Most often she ended up at the barn, brushing Shannon and talking to him. The horse was much past his prime now, but he was still the gentleman he had always been, and still his mistress's pride. His coat was still dark, rich amber, but his mane was showing threads of white amid the chocolate brown. She would brush him and rub him until he shone and then put her arms around his neck, talking to him as though he knew every word she said. Indeed, at times he seemed to. He nuzzled her and poked his head under her hand until she stroked his velvet nose. At last, she would say good night, and he whinnied to her as she left for the house.

She was really more happy with Jack gone than when he was home. Lately, his rages had become increasingly vicious and hateful, and she sometimes feared that he would really hurt her, as Patrick had warned so long ago. She didn't fear for the children. Ruby had a way with her father, and Matthew never seemed to be around during the fighting. He had his own world that he preferred to life at the big house. John Paul was the apple of his father's eye and never heard a harsh word from Jack. No, it was she who brought out his wrath; she who seemed to infuriate him with her obstinance and opposition to his will. Yet, she knew that if she were to back down, he would keep pushing until she was once more completely subdued. Her only hope for his respect was to stand her ground and endure his anger. Afterwards, he sometimes tried to make amends in his own way. She was becoming convinced that when the rages hit him, he actually lost his reason.

It was a dark, windy night. A storm was brewing, and all the hatches had been battened down around the ranch. John Paul had taken forever going to sleep. He had had a million questions about his Daddy and where the wind came from. At last, he was laid quietly in his bed. Charlotte lay down in her room. She listened contentedly to the wind battering the house, feeling perfectly safe and cozy in her home. She had put in a full day working around the ranch, and she was pleasantly weary. Quickly she dropped into a deep sleep.

Sometime in the early morning, she awakened. At first, she couldn't understand what had caused her to wake. She was usually a sound sleeper once she had drifted off. She lay there for awhile, half asleep, half awake, but too tired to want to identify the little noises she half-way heard. The moon was streaming in her window and she watched the patterns of the tree leaves in her room. Then she heard the noises again. Was it mice scurrying about in the closet? Then she heard the scraping again and low, reverberating sounds. It was coming from the direction of Ruby's room down the hall. Suspicion began filling her head. She had long been concerned that Ruby would go beyond flirtation and deceive them with the boys.

It was very dark in the hallway. She stopped to listen. After what seemed an interminable time, the scraping sounds began again, and this time she was certain she heard whispering. Charlotte moved quietly down the hall, avoiding the loose boards in the middle of the floor. She came to Ruby's room but heard nothing at all. Then the creaking began again. It was coming from Matt's old room at the end of the hall. She was sick inside, sure that Ruby had brought a boy into her house. What should she do? The girl had to be stopped, and steps taken to correct this tendency of hers. Charlotte moved toward the door. She stood outside, in the darkness of the hallway, quieting her stomach and steeling herself for a confrontation with her daughter. Then she opened the door.

The moonlight was streaming through the west window here, just as it was in her room. It highlighted Jack's black head bent over a dark-haired girl who was sitting on the edge of the bed with a bottle in her hand. It took a few moments for Charlotte to recover from her shock and realize that the girl was not Ruby. Jack had brought his harlot into her home! She shook her head as if trying to clear away the film. Jack looked up at her, grinning in drunken glee. The girl didn't move.

"Oh, Charly," he whispered, "it's you. We were trying to be quiet. Didn't want to wake the whole house up, you know." His finger went to his lips theatrically, and he turned quickly to Connie, knocking the bottle from her hand. As it broke on the floor, he hissed at Connie "Shh, don't cha know we gotta be quiet?"

Charlotte began turning to ice inside. "What is she doing in my house?"

"Aww, Charly, can't cha be more hoispit . . . hospit . . . hospitable? I invited her. She's our guest."

"Get her out of here, Jack. I'll burn the place down before I'll ever have her in my house."

Charlotte's mind was reeling. She was dizzy and the whole, weird scene was surrealistic. She had had one place of refuge from the pain. One place of security, of comfort, a place to call her own, her home. It was her one shelter against the pain of her husband's unfaithfulness. She had almost closed her mind to his other life, whatever it was. Here she could cope with whatever she had to deal with. This was all that mattered. It was her world, and now this woman was not content with taking her husband, she also had to violate her home, her security. Inexplicably, Charlotte felt dirty as though she had been touched in the core of her being with awful, dirty fingers!

Jack's eyes narrowed dangerously, and he looked like a black panther ready to spring. "Oh, so you think you have any say so over who I bring to my house? Let me tell you, Miss High and Mighty, you are here by my say-so. I give you everything you've got, your fancy clothes, your food, even your children, and I can take 'em back whenever I want. You got nothin' to say about it. Now Connie here, she is just what I like. She don't put on any airs. She ain't got no 'secret souls' that I can't come near."

Jack began to circle Charlotte, and she turned slowly, facing him. "You think you're so much above me, don't you?" he growled. "I never told any woman I loved her but you, and you just laughed. I don't crawl, Charly. I don't crawl for anybody, but you'll crawl before I'm done. Get down there."

She tried to back out the door, but he had reached it first and shut it, barring it with a chair. "Jack, you're drunk. Go on now, send her home and we'll talk tomorrow."

"No, by damn, we'll talk right now. I'm sick to death of fighting a woman to see who's the head of this house. You're gonna admit who's head of this house, who it belongs to, and who has the say-so over it all. Now get down on your knees!"

She stood still and held his gaze. On the floor by the bed, Connie had cut her hand on the bottle glass and began crying softly. Charlotte could see the girl was dressed only in a light chemise. Charlotte felt her own anger rising despite her reason that told her she had to be careful of Jack. He seemed to be swallowed up in his frustrations, his hatred of something he could not possess—her. Her instinct told her that tonight, of all the times they had fought, he was really dangerous. She had to stay calm if she was to escape him. But when she looked at Connie, drunken, flaunting her relationship with Jack, the old indignation at having to compete for her husband came back to Charlotte. The stench of the liquor, the fear of Jack and the anger at being so shamed came rising up in Charlotte's throat.

Before she could move, Jack's hand flashed out and grabbed her arm, twisting it behind her back. His other hand wound itself in her hair, and he forced her head back, while bringing his own twisted face down to meet her's. He breathed his whiskey breath in her nostrils, "On your knees, Missy! On your knees to your husband, and I wanta hear you cry and say you love him, and you're sorry you don't please him."

"I'll see you in hell first," she said between clenched teeth.

Like a shot, he slapped her. The sting brought tears to her eyes, but she forced them back.

He mistook her resolve. She would never crawl to him as he wanted. She was thoroughly sick to her stomach and enraged at the humiliation he heaped on her. Connie called out to him, "Jack, I wanna go home."

In the second he took his eyes off Charlotte, she gathered herself to fight. With all her strength, she rammed her head into his midriff and knocked the wind out of him. He stumbled backward off balance, roaring in his pain. Before he could recover, she was at the door. Charlotte knew she had to get away now or he would beat her senseless.

She raised the chair and as she brought it down on his head, she cried, "I'll never crawl to you, never!"

Outside the door, Emery and Lily were calling, "What is going on? Mrs. Boughtman, are you all right?" She flung the door open, and they could see Jack unconscious on the floor with blood coming from his temple. Connie had crawled over to him and was crying, "Jack, Jack wake up."

Ruby came running out of her room, but Charlotte caught her before she could get to the doorway and shoved her back in her room. "Keep her in there, Lily. I've got to get away now, before he wakes. He'll kill me for sure this time."

Straight down the stairs she ran—no time to dress, to take anything. How long Jack would be out she couldn't guess, but she knew he would be murderous when he awoke. Emery ran beside her and helped her saddle Shannon. Just before she rode off, he flung a cloak around her shoulders, and she was gone into the night.

CHAPTER 9

Charlotte Boughtman had gone as far as she could go. She was too numb to even guide Shannon. For almost an hour, he had been walking doggedly along, facing into the wind, the rain lashing his face, with his mistress bent over his neck. They had left civilization behind hours ago and were entering the wilderness of Iowa. Now there was just a trail, made, perhaps, by some trapper, and the big chestnut horse kept plodding along. After a few miles more, he came at last to a weary stop at a fork in the trail. Whinnying softly to his mistress, he raised his head slightly.

The movement was enough to arouse her. She had been more in shock than asleep. She raised up and looked blankly around her. She had no idea where she was. The furthest she could see in any direction was a few yards. All about her were towering, dripping trees and undergrowth so dense only the rabbits knew their way. She saw why Shannon had stopped. The fork in the trail offered no promise of what was to come further on. Both ways seemed to lead into more forest. She was too drained to care anymore where she was or where she might go. There was nowhere to go. It was done. She had left her home, her family, and civilization. The only thing she had now was her old, beloved friend, Shannon. She slid down off his back, and he followed her beneath a great elm tree. The lightning had long since gone, though it had flashed wildly, repeatedly cracking the sky for the first terrifying hour of her ride.

Any moment she had expected Jack to ride her down, like a hunter after a fox. So she had headed Shannon west for a ways, then south, realizing there was no help for her in Montrose. After ten miles, she had turned west again. Now the world had passed away, and here she was, the last being on earth, at the last place on the edge of the earth, and all she had was her pain. The rain was still coming down hard. She had long ago been drenched, and she sat now, on the ground, on the tree roots, her head drooping on her chest, Shannon at rest beside her.

She sat there for a long time, awake but unaware of anything except her own wracked soul. The wretched life she had endured for the past sixteen years, the eternal battling with Jack, the beatings, the humiliation, the fear of his rages, the last few terrifying moments when she had struck blood from his head—it all came crashing in, smothering her with grief so exquisite she could hardly move. A terrible weight rolled across her soul and she slipped to her knees. Deep within the torn, frightened reaches of her heart, a cry began. It grew and grew, suffocating her as it came. All at once, it broke like a huge wave crashing upon the shore, and she could hold on no longer. The sound came tearing out of her, ripping her apart as it came.

She threw her head back and screamed her pain out into the night. "Oh, God, how can you be so cruel? How can you? How can you? I have hung on until I can't anymore. You have left me nothing to hang onto. Let me die! Oh, God, if you are reallythere, if you have any mercy in your heart, let me die. Let me die! Let me die!"

Her voice rose to a wail, like an animal suffering in the steel jaws of a trap. She sat back on her knees, the tears gushing out with the streamlets of rain on her face. Over and over she cried His name, strangling on the word, begging to die, as she had never begged for anything before. At long last, when still she lived and breathed, indignation began to mount. Anger came and leaped upon her.

Finally she raised her fist to the black and weeping sky, shaking her head along with her fist, crying bitterly, "Ye won't even give me that!" She screamed again, pain at the very fact of living ripping at her.

"Ye are a cruel God. Ye must delight in me torment. Ye must glory in this damnation. Me whole life has been a damnation, and ye have no pity. Have ye no lightning bolts to crash upon me head? Do ye not hear me blaspheme? Kill me, kill me! I curse ye and yere earth and the breath that gives me life!" Her voice rose and rose, while she called down death and damnation for her blasphemy.

Then the blackness overtook her, swiftly, mercilessly, and she sank to the ground in the grip of a darkness so black she believed that God had granted her desire. Shannon had been fidgeting nervously, but as she sank down, he grew wild with fright, rearing, snorting, whinnying, and calling to her. He came dancing in close to her, only to jump back as though shocked as his nose touched her shoulder. He reared, pawing the rain with his hooves, striking the earth and pawing again, frightened by a sense of danger.

Still she lay as though dead, scarcely breathing, yet conscious in her black hell. This was a power more terrible and strong than she had ever experienced before. It was much like Jack's black hole of anger. She could not have struggled if she had wanted to, for she had no resources to combat it. She was enveloped in utter, pitch blackness, dragged down and down to hell, held powerless by a force so evil she had no name for the terror of it. She had no will or volition of her own anymore, and the power had a strangle hold on her soul, slowly crushing the life from her.

After a long, terrifying time, a dream came to her through the darkness. She was back at the Governor's ball, sweeping through the ballroom in her blue gown, dipping and swaying with the music. And then she saw again the dark-haired stranger, and this time he came to her and moved into her arms, holding her gently, firmly, guiding her through the crowds, out the huge doors and across the grassy meadow. The moon was behind his head and the wind caressed them both. She felt nothing beneath her feet, yet the meadow and trees and the stream went fleeting by beneath them. She only knew the warmth of his arms, the strength of his body near hers. He spoke her name over and over, like a caress. Love and peace swept over her, supporting her, lifting her up, and freeing her from the nightmare of hell that had been so close. Here was safety; here was refuge. Here was love! His lips brushed her forehead, her cheek, and firmly, tenderly settled on her own lips, and she gave herself to a deep tide of love that drew her into warm, dark oblivion.

She was not to know until many, many years later that she was pulled back from the mouth of hell by this man, dark-haired and gentle of countenance, and dressed all in white. He came and stood over her body, lighting up the night with his glory, exercising all his heavenly power to dispel the dark grip that held her soul. He stood above her for a long time, silently battling the darkness with the force of his goodness and love. At last she began to breathe normally again. Released from the terrifying grip of evil, and drained of all ability to move, she drifted off into a stupor-type sleep. The glorious being looked down on her with tender eyes, the eyes of a poet or a saint, dark and deep as the ocean, covering her with a sweet, protective love. He would have touched her if he could, but his body was purer and more refined than hers, and he could only look at her, pity her agony and leave her a dream of love.

A few hours later, Michael Gailbraith found her unconscious on the ground beside the tree, with a chestnut horse lying protectively beside her.

Michael Gailbraith was on his way to Lansing, Iowa, with a mule loaded down with furs. He had been out in the wilderness for four months now, and he was getting hungry for some decent meals and one of those soft beds the city folk sleep on. Not that he would ever trade his stars and streams for their houses or beds; he loved his free life too much. Still, every now and again it was a treat to sink down into a feather mattress. It was like the soft, fleshy arms of a good woman.

His woman was long since dead. She had died eleven years before in her second childbed. The baby was breach, and there was no one to help her when her time came. She struggled and pushed for hours in the little cabin on the outskirts of the wilderness. When a traveling minister found her two days later, she and the unborn baby were both dead. One little two-year-old boy sat, silent and frightened, in a corner of the cold room. Michael had come a week and a half later to find his house empty and his one son living in the nearest town with Mrs. Palmerory.

Since that time, Michael had never taken another wife. He would not subject another woman to the cruelty of the wilderness, and he had never met one for whom he would change his life.

Gailbraith was a Scotsman. He had immigrated from the coal mines of Scotland to a new land where they said a man could breathe pure air and live like a king. It had all come true for him. The air he breathed had no coal dust; it was as sweet and clear as the streams he fished and drank from. He had his freedom. He had green hills and plenty to eat, a good hot bath once in a while and plenty of time to make friends with God's creatures. Not that he was what you'd call a religious man. You'd never get him in a church, and the technicalities of religious doctrine did not interest him. What interested him was the way the birds called to each other in the early morning mist, the way a male badger would stay with his female mate when she was trapped, the way the sunset never changed in beauty but changed it's picture every day. God's world was his love, and sometimes he talked to Him about it.

"I like the way ye made all those fish to taste just a little bit different," he would say to the Creator. "Twere a pleasin' and thoughtful thing ye done."

"Tis sorry I be about that raccoon. I niver meant for yere animals to suffer. I killed it quick as iver I could," he apologized, after the wolves had half eaten one of his trapped animals.

For a man as soft-hearted as Michael Gailbraith, it was rather a strange occupation he had. Often he begged an animal's pardon when he had to kill it. Still, it was the only life he was adapted to, or wanted. He was not affected by the loneliness. Singing his Scottish songs, talking to his animal friends and whistling back and forth to the birds, he little felt the loneliness of his situation. In fact, he was frequently surprised by the queries of an occasional townsman as to how he survived the loneliness.

The only thing he did miss was his woman. He had tried a time or two to use one of the easy women that every town had. It was just unnatural to him. He got no pleasure from it, and it only made him hungrier for a woman of his own. What he really missed was the companionship of a woman. For he could talk to women. They understood the love he had for the natural beauty of the wilderness. The tender side of Michael's nature yearned for an outlet, one that he never had in the company of men. At first glance many of the genteel women of the town were afraid of him, but if they once spoke with him all fears were banished. Michael could no more hide his gentle nature than a sunbeam could hide a rainbow.

Victoria Palmerory, Michael's self-appointed mother, was a spic-and-span housekeeper. Her sheets always smelled of a fresh breeze, and her kitchen of hot tea. She had the patience of a saint and the single-mindedness of a trail boss. It was to her that Gailbraith brought Charlotte, weak and coughing and only occasionally conscious. It was readily apparent she had pneumonia. Her fever was rampant, and her breathing shallow and difficult.

He rode in with Charlotte's Shannon and his mule in tow, the girl slumped over in front of him on his horse. Victoria was just biting into a breakfast muffin when he arrived and deposited Charlotte in the rocker of her front room.

The first thing she asked was, "Where did you get that?"

The first thing he asked was, "Where's ma' boy?"

"Joshiah's over to Vanderhousen's. Decided he wanted to apprentice out to a carpenter."

"Found her at the fork a few miles back. Looks pretty bad off. Could ye be a holpin' her?"

"Better get Doc over here." They laid Charlotte in the bedroom on the second floor, the one with the windows facing east and the yellow organdy curtains. She wouldn't suffer the girl to be laid on the bed until Gailbraith had gone out and she had stripped her of her filthy, sodden nightgown. Privately she clucked to herself over the character of a woman who would go out in the wilderness with just a nightgown and a cloak. She tugged and pulled one of her own substantial, voluminous nightgowns on the young stranger and hefted her onto the bed.

Michael returned with the town doctor shortly after Charlotte was properly dressed for sleeping between Victoria Palmerory's sheets. The doctor, a combination tooth-puller, veterinarian, physician and store owner did his job quickly. It didn't take much expertise to see the woman was in the full grip of pneumonia. After some instruction on mustard plaster to relieve the congestion, and herb teas to bring the fever down, he told them the best they could do for her was to pray.

As she closed the door on the good doctor, Victoria muttered, "Hah, I could have prescribed as much!"

While the talk flew between Mrs. Palmerory and Doc Wiggins, Michael looked down on the woman he had brought in. What he saw was a woman in her thirties with dark reddish hair—wild and matted now—and a generous sprinkling of freckles standing out against pale skin of her cheeks and forehead. Her mouth was delicate, and her eyebrows were drawn up as though she were in pain even in her sleep, and there was something so sad about her face that he felt about her as he often did his animals caught in a steel trap. If they had been alone, he might have whispered an apology for her hurts and for the life that had so pained her.

It was three weeks before Charlotte was completely out of the woods. At times, she was conscious, though always silent. Other times, she slept fitfully. That was the only time she ever talked. Michael spent several hours a day, generally in the early evening when she seemed most restless, sitting by the window and the organdy curtains. What he heard usually caused him to shake his head. Most of her ravings were directed to God. Over and over, she asked Him to let her die. He heard the name John Paul many, many times over, but he never heard the name Jack.

Mrs. Palmerory was completely faithful in her care of Charlotte and cheerful in the face of her silence. After three weeks, she came to take the breakfast tray and announced, "Now we will get up." Charlotte shook her head. Victoria paid no attention whatsoever and began helping her out of bed as though it were Charlotte's own request. At that Charlotte protested.

"I'm not ready. I don't think I can walk."

"Nonsense! Land-a-day, my dear, you've been a-bed for nigh onto a month now. Your limbs'll go rusty less you walk about a bit. Besides, it's the cheerfullest day you ever saw. Pure tonic for a body."

There was nothing for Charlotte to do but go along. She was too weak to fight it. Leaning heavily on the older woman as she barely shuffled a few feet around the room, she failed to elicit any sympathy from Victoria.

"See, I told you the walk'd do you good. Strong as a horse. I knew it the minute I laid eyes on you."

She walked Charlotte through the door into the other bedroom but thought better of trying the stairs. "Enough's enough. Plenty of time to get you downstairs. You'll be meddling in my kitchen soon enough I 'spect."

That afternoon Michael came and talked to Charlotte for a very long time. He even took his meal in the room with her. All the memories he had of Scotland he painted before her. She rarely spoke but watched him and listened attentively. He managed to coax some information from her.

"My father was an immigrant, too," she said. "From Ireland, actually. He used to talk a good deal about the old country."

"What were yere father's name?" Michael queried.

She hesitated. "Patrick O'Neill."

"Ahh, well, O'Neill! A vera fine name for an Irishman. What be ye called?"

She paused a little longer this time, and he began to think she was hiding in her silence again. At last she answered, "Annie. Annie O'Neill."

Happily he went on talking, sometimes whistling his little Scottish songs, until at last she drifted off to sleep.

Those days were pleasant to Charlotte, days she had never dreamed could be. There was no yelling, no threats, no confrontations, no dark, angry temper to deal with. Indeed, she had forgotten such uncomplicated people really lived. She had lived for so long with dark clouds forever breaking about her, she had not realized how refreshing and renewing it could be to live with kindness and cheerfulness. Under the influence of Victoria's relentless optimism, Charlotte began to allow her spirit to climb. In fact, it would be very difficult to resist the good-natured cheerfulness of Victoria Palmerory. Her first word in the morning, rainy or sunny, was unfailingly, "Now, isn't this a marvelous day the good Lord has sent us?"

After she was able to get up and about sufficiently, Charlotte began to insist on helping around the big rooming house. Never in her life had she willingly done housework, preferring even the hardest outdoor labor to it, but now she proved a meticulous cleaner. Soon she was taking over many of the jobs of polishing, dusting, gardening and keeping the guest rooms spotless. Her work was precise enough for that spit-and-polish lady, Mrs. Palmerory. Charlotte knew she could not stay on here forever living on Victoria's good nature, so she endeavored to make herself useful until she could decide what to do next.

She had not even considered what to do with herself should she be so unfortunate as to live. She had expected to die. She had no plans, no money, no resources and no future. If she stayed in these parts, someday Jack might find her, and that she could not endure. So she began to ask questions about such places as St. Louis and New Orleans.

Poor Victoria! She hardly knew whether to enjoy the mystery of Annie O'Neill, or to be immensely irritated over not being invited into her secrets. She never tired of trying to ferret out new pieces of information about her identity and past. She suspected Annie was not her real name. She made up fantastic tales in private about the background of the young friend. And Victoria and Charlotte became friends—warm, unassuming friends.

Victoria Palmerory was always the first one through the doorway of the little church down by the road leading out of town, and she sang with more gusto than anyone else (though occasionally off-key). All good things she attributed to the "good Lord"; all misfortune to the unpredictability of life.

Once she told Charlotte, "Now Michael, he sets a great store by that boy of his. Still blames himself, he does, for his wife's death. Thinks he could have saved her if he hadn't been gone. But you know things like that are bound to happen in life. Women die all the time in childbirth. Isn't anything anybody can do. It's just natural. We ought to be grateful for the good years the Lord gives us."

Victoria tried several times to get Charlotte to go with her to "meeting", but her young friend declined politely. At other times, Victoria would try to discuss the Good Book, and Charlotte never offered any commentary. In exasperation, she once asked Charlotte, "Haven't you ever been educated to God's way, Annie O'Neill?"

Charlotte had answered, bitterly, "Oh yes, He did a fine job educating me or 'learning me good' as my Papa used to say, and I almost died from such an education."

Victoria could hear the hurt in her voice and asked more gently, "But what did you learn, my dear?"

Charlotte laughed shortly, "That He doesn't care, if He is there at all, so it is no use going to Him for anything. If He takes any notice of us, He looks on from a distance and all our pleadings fall on deaf ears."

Victoria shook her head sadly, "No, you're wrong. He isn't deaf, and He isn't at a distance. He's close by and He cares more than you can know. Sometimes, I think He suffers with our mistakes and sorrows more than we do ourselves. On rainy days, I often think that He is weeping for someone's pain. Myself, I haven't had much pain. Oh, a few years back, I lost Mr. Palmerory, as good a man as any woman would want. And I used to worry a lot that we never had any babies. I'd have liked some babes. But I'm not used to brooding much. I just take each day as it comes and thank the good Lord for the sunshine and thank Him for the rain; flowers need both, you know, and I hope someday to be a beautiful, bright peony in the Lord's garden."

Michael had little more knowledge about Charlotte than did Mrs. Palmerory, and frankly, he didn't care about her past. He had grown to respect and admire her. She suffered her illness and misery in silence, and he guessed that she was suffering inwardly even more. As she overcame the initial onslaught of the pneumonia, he reluctantly gave up sitting by her bedside at night until she fell asleep. The last time he sat with her, she drifted off to sleep several times, and then seemed to go each time into a nightmare which would wake her up within a few minutes. She would lie stiff and wide-eyed with terror while he soothed her as he would have a sparrow with a broken wing. He moved his chair close to her bedside and sat with her until late into the night, stroking her cheek very gently, soothing her by his gentle protection.

Michael was a wonder to Charlotte. He looked like such a wild man, with his bushy, blond beard and frontier clothing, but he was the gentlest man she had ever known. Unfailingly a gentleman, he never pressed her to talk about ked into his blue eyes, she wondered what her life would have been like if she had married such a man instead of Jack.

As her health improved they spent hours sitting in Victorial's swing, shelling peas for dinner, or walking by the banks of flowers that grew around the house. When at last he left her, he went straight to the woods and back to his beloved streams and paths. He stopped to say good-bye to no one. He had too much on his mind. After a few weeks in the forest, he made up his mind and, just as suddenly, he struck out for town again.

Charlotte was in the garden, weeding Victoria's flowers, when he came. She turned to him, pleased to see him again, and smiled into his clear blue eyes.

"Annie," he said. "Annie O'Neill, will ye be ma woman?"

Charlotte jumped like a doe shot. He took her by the shoulders and turned her to face him. "I know I bein't a good ketch. A wild mon I am, livin' with the wild animals, gone for months at a time. And I must confess to ye I can na' change me life. But I need a woman, and I took a likin' to ye the vera first time I saw ye. So I'm askin', will ye be ma woman?"

She watched his face, trying to sort things out in her mind. In the past month, she had been doing a lot of thinking. She had not decided anything except that she must somehow get John Paul away from Jack and go somewhere far away. Her discreet inquiries as to St. Louis and New Orleans simply convinced her they were out of reach. Ruby and Matt she could never influence away from their father. His hold was too strong on them. But she would not allow little John Paul to be ruined by Jack. There was no possibility of going back to Jack. If she did, she knew it would only be a matter of time before he would kill her in a fit of rage. If she submitted, to avoid his rage, she would eventually die of depression and neglect.

Charlotte gazed into Michael's calm, blue eyes, wondering if he would want her if she told him the truth, and wondering if she wanted another man, even such a gentle man as Gailbraith. For a moment, she felt a bitterness that she had not known him when she was sixteen years old. Life would have been very different for her. Could she marry him now? She was already married. It would be bigamy. Could she deceive him? No! That was clear. He had been too kind to her.

He was waiting for her answer. She looked at him and smiled for the first time in a month. He was dressed in the soft skins of the animals he trapped. His shoes were high boots strapped to his calves by rawhide strips. His beard was a soft scrub-brush on his cheeks. His hair was almost white from exposure to the sun and wind, and his skin was a honey brown toughened to leather. He looked like the wild man he called himself, yet an emanation from within conveyed a gentleness beneath that rough exterior.

"Mr. Gailbraith, could we perhaps sit on the grass in the shade for a moment? I'm feeling rather faint and tired."

"Ach, I am a wild mon! Niver stoppin' to ask how ye be a feelin'." He guided her carefully, a hand under her elbow, chiding her gently for doing too much too soon.

After considering her words a moment, she began, "Mr. Gailbraith, I'm not sure I have ever thanked you properly, nor ever could, really, for all you've done for me. Instead, I have simply taken your charity and given nothing in return. Now you flatter me with a proposal." She paused, then said gravely, "I do suppose your proposal is one of marriage?"

Embarrassed, he hastened to assure her it was.

"I would like nothing more than to give back some measure of all you've given me. To be fair, however, I must confess that you know very little about me."

Michael began to protest. The last thing he wanted to hear was a confession about a sordid past. He really didn't care what sort of a past she had. He only wanted her future.

"No, really, there are some things you must know." She sighed heavily. "To begin with, I am not Annie O'Neill. That is my sister. My true name is Charlotte O'Neill Boughtman. I am a married woman." He didn't make a sound, but his blue eyes filled with disappointment and finally looked away. "My husband is a violent man, and, when you found me in the forest, I had run away from him. In order to protect myself from him, I struck him and brought blood to his head. So for all I know, I could be a murderer, though I could hardly be so lucky," she said bitterly.

"If I had no husband and could think of marrying, there is no one on earth I would rather trust myself to than you, Michael Gailbraith." She put her hand softly on his sinewy, gnarled one. "You are a true gentleman."

He waved her compliment away, "Ach, I am no gentleman, lass. I would na' have ye fooled. A wild, trappin' mon I am."

There was silence between them for a few minutes. Then he spoke, looking down at his hands where one of hers still lay softly. "Ye are not goin' back to him, lass?"

She shook her head.

"Well then, could ye be for goin' with me? I've niver had a woman in the wilderness afore, but I promise I could take keer o' ye. I would take ye where no one could find ye but the deer and me."

"What about the wolves?" she asked smiling at the romantic picture he was trying to paint.

But he waved them away, "Wolves—shadows—anyway, I would be with ye all the time. No lass, sounds like ye would be safer with the wolves and Indians than with yere husband."

She nodded her head in agreement.

"Do you still want me then?" she asked.

He grew pink underneath the leathery skin. Humbly he answered, "Aye. That I do. Ye are a handsome woman, Charlotte O'Neill Boughtman. Yet for all that, I'd not be a takin' another mon's wife, unless he was na' deservin' of her."

"I shall have to have some time to consider it, Michael. There are many things to ponder. I would be leaving my children if I went with you. I have two older ones and a little boy just two years old. I must consider what to do about him."

"Is he the John Paul I hear'd ye ravin' so about."

"Yes."

"I'll give ye all the time ye need."

The next day he asked her if she had made a decision. No, she hadn't. The day after that, he tipped his hat to her and asked again. No, she hadn't. On the third day he fidgeted all morning around Mrs. Palmerory's house, trying not to ask. Finally Victoria said, disgustedly, "Oh, for heaven's sake, go ask her, you ninny."

He looked at her in big blue-eyed surprise. "How did ye know?"

She hrumphed like a big bull frog. "Anyone'd know what you got on your mind. You go and peek at her every few minutes, twist your fur cap in your hands and scrub your beard with your fingers. You sit down with a sigh like the winter wind. How do you 'spect any self-respecting busybody like me not to know?"

"Do ye think she will marry the likes o' me?" he asked anxiously.

"Well, if she don't, she's the Lord's own fool!"

After the fourth day of no decision, Michael told her that if it was living in the wilderness that was scaring her, he would leave it and find work in the town. Charlotte was touched, and kissed his cheek.

That night she deliberated all night. The past two months had been like a happy dream. Could she be so fortunate to be rid of Jack? Could she really go far enough away that he would never find her? It seemed too good to be true. After so many years of torture and misery, was there really a chance for a happy normal life for her with a man who so obviously adored her? If only John Paul were not left behind with Jack. She should have taken him that night. Finally, she decided that if anyone could help her to steal John Paul away, it would be Gailbraith with his animal cunning and silent footsteps.

That she was considering adultery or bigamy didn't really concern her. Those kinds of moralistic words had ceased to have any meaning for her, as she had decided that God was either mankind's desperate imagination or, worse yet, an unconcerned creator. All she knew was that Michael Gailbraith was a kind, tender man who was ready to love and protect her, and she was so tired of fighting life, of trying to be strong and battling the tempest alone. She wanted to curl up in his pocket like a soft, little rabbit. He would stroke her and feed her, and she would be safe. If only she could get John Paul, life might have some meaning yet.

The next morning, when she came down to breakfast, Michael was not there. Mrs. Palmerory said that he had gone to visit Josiah, his son, for the day. It was while Charlotte was washing the dishes in the sunny kitchen that a young man knocked at the door.

Mrs. Palmerory answered it and saw a seventeen-year-old boy with dark brown hair, a sprinkling of freckles and hazel eyes. He removed his hat and asked if Mrs. Charlotte Boughtman was staying there.

"No," Victoria said, curiously. "And who wants to know?"

"Her son, Matt. If she hasn't forgotten me," he said somewhat bitterly.

"Well, there's no one here by that name," Victoria started to close the door.

"The man over at the general store said you had a guest answering my mother's description. And I know her horse, a big chestnut, is at the livery stable."

"What does your mother look like?"

"She is about thirty-four years old with red-gold hair, freckles like mine, and about this high," and he measured off a little better than five feet. It was almost a foot below his own head.

"The only lady guest I have is named O'Neill."

"That's her, Charlotte O'Neill Boughtman, my mother, and she come on her horse, Shannon."

"She did not," Victoria said smartly. "She come draped over the shoulder of Michael Gailbraith, who found her sick unto death in the woods. You the devil that drove her out?" She peered accusingly at him.

"Uhh, no ma'am," he stammered, not knowing what to think. He had not thought about her plight, running away on a stormy night. "Anyway, I've got here a letter for her from my father."

She took the piece of paper he was extending to her. "Hrumph, well, I'll go ask 'Miss O'Neill' if this belongs to her."

She found Charlotte wiping down the table and window sills.

"Annie, there's a young man to see you. Claims you're his mother, and says this here letter is for you."

Charlotte felt her stomach go sick. She knew it was Matt, and that meant Jack also knew where she was. Her first instinct was to run out the back door, to find Michael and get him to take her away as quickly as possible. She had a clear, mental picture of Jack riding her down into the dust on his big, bay stallion.

"Is there anyone with him?" she asked weakly.

"Not that I can tell."

She took the letter and slowly she opened it. She knew it would be from Jack. It was simple.

Charly,

If you want to see John Paul again, you'd better get home. The paint mare kicked him the other day when he was fooling around with her colt. He's bad off, and we don't expect him to last long.

Jack

Now she ran out to Matt, threw open the door and grasped his arms.

"Matt, is the baby all right? I mean, how long ago did this happen? Where was he hit?"

"The mare kicked him right in the head. I told him to quit messing with her colt, but he's so stubborn, he wouldn't mind me, and the next thing I knew she knocked him against the wall of the stable. I was lucky to get him out before she could paw him to death."

"When, Matt, when?"

"Must have been about a week ago, but we've been looking for you ever since you run off," he said darkly.

"Come on in and wait. I'll be ready in a few minutes."

Victoria followed her flight up the stairs. "You going?" she asked. "Just like that, you're going? What about Michael?"

"I don't know," Charlotte answered, packing the few clothes she had acquired. "Tell him my baby is hurt bad and I've got to go. He's been a dear, and so have you," she turned to Victoria and clasped her hands for a moment. "You've both been so wonderful to me. I've been so happy with you, I wish I could stay always. Perhaps I'll come back one day, depending on what happens to my baby. Tell Michael I live outside Montrose, Iowa, if he ever wants to find me. Victoria, I'm sorry I had to deceive you. I didn't want my husband to find me; our life together has been hell. The only thing that means anything to me is my baby. I have to go back, can you understand that?"

Victoria's trim and tidy mind was in a whirl. But she nodded, "Yes, a woman hasn't ever got love for any man like she has for her babies. I understand."

Charlotte kissed her on the cheek, thanked her once more for all that she had done for her, promised fervently to write to her, hugged the older woman quick and hard, and ran down the stairs.

"Come on Matt, we haven't any time to lose."

Once more on Shannon's back, riding headlong into the terror she had escaped, her head was filled with both fear and anxiety lest John Paul die and leave her heart totally empty.

Matt asked her once as they rode side by side, "If it had been me dying would you have hurried back?"

She looked at him, tall, manly almost, but still a boy in his heart, and she could hear sadness in his question. Love flooded over her, she reached out to him. "Yes, Matthew, I would have come as quickly as I could. You were my firstborn, my first joy. I wish I knew how to wipe away these last few years and make things right with you."

"Mama," he asked timidly, "Why did you hurt Papa and then run off like that?"

She hesitated, trying to decide how much of the truth to tell him. "We had another fight, a bad one, and I was afraid he would kill me if I hadn't stopped him. How badly did I hurt him?"

"He was in bed for a couple of days. The doc said to take it easy so his cut wouldn't open back up." He grew dark and moody again. Jack had told him that he had caught Charlotte with another man, and that she had hit him with the chair in order to escape with her lover. He had believed that lie, as always, yet now, as he looked at his mother, instinct told him she was not covering up. Matt was confused.

It only took them five hours of hard riding, but it seemed as though she had been in a whole different world—a world with flowers and sunshine and tranquillity. Now she was riding back into hell. Her one chance to escape was gone and she had given it up of her own free will. She patted Shannon's neck. At least she still had her old friend—and her baby, she hoped.

CHAPTER 10

Montrose was not much different than when Johnny had left it some ten years before. Hector's old blacksmith shop was a livery stable, and Miss Ann still ran her boarding house. Now, however, she had a permanent boarder. The Reverend Henderson had married her a few years before and moved into town with her. It was a perfect arrangement. She could work and he could preach.

Johnny had left Faith and his family, now five children—all boys—when Brigham Young had called him on another mission. He had been in Stake Conference one beautiful, spring Sunday. Orson Hyde was conducting, and in that meeting had called four men out of the audience, announcing that they had been called on missions. It was the first Johnny had heard of it. Faith had wept that night in his arms, and after she had gone to sleep, John had held her softness, caressed her hair and wept also. But he went.

He was headed to eastern Canada and his companion was an old friend, Hector Thornby. Johnny had been right. Hector was considered a valuable addition to any community out in Utah, and he belonged to several due to the fact that he had four wives and kept them all in different towns. After so much celibacy, Hector had really taken to married life and wanted as much of it as he could get! Now he and Johnny were on their way to their mission and stopped off in Montrose for old times' sake.

After visiting with the Reverend, and stopping by Johnny's old place, O'Neill rode on to Charlotte's. Hector remained behind in Montrose. Johnny wanted to see her alone. He could hardly wait to surprise her! How should he present himself? At the front door like a proper stranger? Climbing in through a window and surprising her? Maybe catching her outside? He still hadn't decided when he came in sight of the Boughtman ranch. He rode up to the gate and watched. Was Jack home? Why was everything so still and quiet. No work was being done today; animals were lazily munching on hay, no workmen bustling to and fro. It was strange. He waited for awhile; then curiosity got the better of him.

John Patrick knocked at the front door, and it was quickly opened by the Negro woman, Sophie.

"Is Mrs. Boughtman home?" he asked, his hat in his hands.

She put her hand to her mouth. Her eyes grew large and rolled once, and she fled, leaving Johnny at the doorway.

It was only a minute before Jack appeared. He looked ragged and older— much older than when Johnny had seen him last. Now he was not arrogant. He looked as though he had been through a long illness, unshaven and disheveled, his eyes bloodshot and heavy.

"She's not here. What do you want?"

Just like that. No hello, no acknowledgment of family. It might have been just yesterday, instead of ten years ago, that Johnny had sat on his horse and demanded to see his sister.

"What is wrong here? Where is Charlotte? I can feel there is something wrong. Is she sick?"

Jack gave a short, harsh laugh. "I don't know where the hell she is. Been looking for her for a month. She's gone. Thinks she left me for good I 'spect. She'll be sorry." His tone was not threatening, more like gloating.

"What happened?" Johnny asked. He could feel a strangeness in the air.

"Why the hell should I tell you? Git gone. She's not here and probably never coming back."

Jack had no intention of telling O'Neill anything about Charlotte or their little boy. He didn't want the Mormon praying over his boy. Nobody was going to touch that little boy but him, especially that holier-than-thou brother of Charlotte's. He gloated inside. Charlotte wouldn't even know her precious brother was here.

Charlotte and Matt rode in just before dark that day. She jumped off Shannon and ran quickly upstairs. Charlotte sat with John Paul in her arms all evening and night. She had had nothing to eat and didn't want anything. Her whole concern was for her baby. He was so beautiful—golden-red hair, long eyelashes, smooth, baby skin delicately browned by the August sun. She could not believe that she would lose him, though he had been in a coma for a whole week now. They poured milk down his throat to keep him alive, but he was wasting away. His head was bandaged where the horse had kicked him. She had not stopped looking at him for hours.

Jack sat across from her in his old armchair, his feet propped up on a footstool. He looked old and tired. He had sat for three days, just the way she was now, holding the boy for hours, afraid to put him down for fear he would die. They had not spoken for what seemed like an eternity.

"Charlotte," he said at last. "I must have been out of my mind that night. I don't know what gets into me sometimes. Sometimes I love you so much I'd cut off my right arm for you, and other times I feel so mean and angry that I just want to hurt you. I know that doesn't make sense. I can't help it. It's like somebody else takes over inside me and starts hating you and everybody else."

She didn't say anything. She neither wanted his apology nor his love. She only wanted John Paul to be all right. But Jack went on talking.

The shadows had overtaken them, and they sat, each one lost in his own world. "I think it all started with Joseph Smith moving his Mormons here. I thought I had gotten away from them, and they had to set up their Celestial camp across the river. You know, I didn't always hate Mormons so. I used to work for one back in Ohio." His voice trailed off for a minute.

"Fact is, I was fool enough to let him get me in the river one day. He dunked me under and said I was baptized. Hell, I wasn't baptized! I just wanted to go to meeting to see all the young girls that were joining. Course they wouldn't none of them have me if I wasn't baptized. So I let him do it. Made everyone happy, and I wasn't none the worse off for it, 'til they started preaching to me about my drinking and an occasional fling with the ladies. Said I had to lay off them both. All on account of Smith said so. Smith made the rules and I was supposed to abide 'em. Well, I couldn't do it. I told ole man Hardy the day I walked off his farm that I didn't care if God himself laid down those rules, I wouldn't live 'em."

Charlotte was listening now, but still silently, afraid to break his reverie. It was all news to her. Of all things, she had never suspected Jack of being baptized Mormon.

He went on, "I finally met their great prophet himself one day and we wrestled. I like to had him when he called off the match. He told me that day that I'd never beat him. Well, I showed him, didn't I?"

He paused a long time. Finally she spoke softly, "Did you?"

"Damnned right, I did. Shot the cussed man right through the heart." He looked up at her. "Does that surprise you?"

"No, I guess I knew all along."

"Sure ya did, Charly. That's what I always liked about you. We was both the same inside. You didn't say anything cause you didn't really care if I shot him or not. You never gave a hang for all them people. Guess I went a little too far, though, when I burned out your Pa. Didn't want to, Charly, I tried not to, but he had been at the head of our list for months, and they was egging me on. Finally I just had to if I wanted to keep their respect."

"What about mine?" she asked quietly.

"Aww, well, I figured you wouldn't find out about it, or wouldn't believe it was really me. I wasn't looking for you to find me out that very night. Then when you came up so all-fired mad. Lord, you was something."

They were both silent then for a long time, and the room was all dark around them. It was more than an hour later when Jack broke the silence by asking, "Are you gonna stay?"

"I don't know," she answered. "A lot depends on what happens to John Paul."

"I know I haven't got any right to ask you, but I wish you'd stay."

She looked away from the baby. "Why?" she asked him. "Why do you want me to stay? We'll just end up hurting each other again."

He stared at his feet, embarrassed. "These last couple of months have been a torture to me. Talk about black. It's been bad, Charly. I've been so sick at heart I would like to have killed myself."

She stared at him for a long time trying to comprehend, trying to unravel his mysterious mind. He was so unfathomable, so unlike Michael. Michael was as clear as a June sky, no sudden storms, nothing hidden. With Jack she always felt something lurking in a dark place. She would have been afraid of him if her life meant anything at all to her. It didn't anymore. Only the boy, John Paul, was of any importance.

She went back to her unwavering study of his face, listening to his shallow breathing. He looked so much like his uncle. John Patrick, where was he now? Happy, no doubt, on his farm in that wild desert out west, a million miles away. She smiled cynically for a brief moment. He had said he would always be there if she needed him. God would tell him, he had said. It was pathetic, almost, the miles that separated their persons and their spirits. She no longer believed there was a God. That night in the forest, hadn't she begged to die? If there was a God, he would have granted that plea, either out of mercy or out of anger at her blasphemy. He did neither. She had felt nothing from heaven. Charlotte shuddered for a moment at what she had felt, that terrifying, destructive force that had held her. It had been beyond words or names, unidentifiable, yet, horribly real. She had been a mouse, dragged by a huge, hideous animal off to a corner to be consumed. She shook herself and tried to think about something else.

Life is the master of desperate ironies. She had wanted to name this child after John Patrick, little realizing he would suffer the same injury his uncle had. That night, so many years ago, came back to her. Johnny had lain on Margaret's and Patrick's bed. She could remember her mother sitting beside him, watching him the way she was now watching her own child. She had sat on the floor on the other side of the bed. When Bishop Logan came, she had felt the first gleam of hope. Then he and Pat had anointed Johnny's head, and his father had pronounced the sealing of the blessing on him. She could still remember the words, "Thou hast a great work before thee, and it is not the desire of God that it should be thwarted. So sleep, my son, and when thou wakest thou shalt be well."

He had had a terrific headache for several days. But, within a week he was up and about, occasionally chasing her around the yard. At times, even many years later, he complained of shooting pains in his head.

Now it was John Paul's turn, and she had no Bishop Logan or Patrick O'Neill to call a blessing from heaven. Indeed, if there was a heaven, it was ignoring her. No matter! Her will would keep her son alive, just as Patrick's had done for his son. Charlotte hugged him to her protectively, trying to give him her own heartbeat, to pull him back with her love. He would live! She would not let him die! All the stubborn force of will she possessed she exercised now for her baby.

It was almost dawn when she felt her baby tremble. His soft, fair form shuddered for a moment and then grew still again. John Paul was dead.

John Paul was dead! Jack tried to take him from her after the first initial disbelief and shock. She clung to him even more ferociously. Jack rode to town and brought back Dr. Stoney. He was a big, bluff man disparaging of any but the most severe hurts, believing that his patients should be up and working as quickly as humanly possible. He was kind to Charlotte, talking a long time about life and death and God's will before trying to take John Paul from her. She still would not give him up, to anyone, for any reason. To her, he was not irrevocably gone. He would come back, she whispered inside. So she clung to him, waiting for when he would return, the breathing start again, and his blue eyes open seeking his mother's face. She would be here. She would not fail him again.

They all thought she would go to sleep soon. Jack knew she must be exhausted. She had ridden all day and sat in that one position ever since. Surely she must sleep soon. At noon she was still sitting in her rocking chair, and John Paul was growing stiff in her arms. There were no tears. She was long past tears. On the parched, burned-over desert of her heart, there were no cooling teardrops. She was empty as a dried up well.

After much consulting, Lily tried her hand. She knelt down beside Charlotte. "Little Johnny always liked that blue suit you made him last spring. Let's take him upstairs and put on his suit. He'd like that."

Charlotte considered for a moment. Yes, that would make him happy. So they did, and when he was dressed she laid him on her bed, and she lay down beside him. In minutes she was deeply asleep. Jack took him away then and set about making a suitable coffin. Each nail he hammered seemed to go straight into his heart. There were few things Jack loved. Now this one was gone.

As far as Charlotte could see in all directions was a long, deserted, parched and desolate plain. She found herself wandering, alone and lost in a sun-bleached desert. The sky above her stretched from eternity to eternity, unbroken, mercilessly hot. The wind was her only companion, and a cruel one it was, bearing down relentlessly with its hot, searing power that dried her very soul. It buffeted her and opposed her every effort. Beside its strength she had no will, no power. The wind came barreling down the long, hot plain carrying dust and the fire of unquenchable thirst. She tried to hide from it, but there was nowhere to hide. The sun and the whistling wind sought the quarry, a tiny, shriveled soul bereft of all resources, and bore down upon her with a consuming will.

Charlotte had thought there could be no worse pain than she had already endured, but oh, how she was wrong! A pain beyond cries, beyond screams, and a terrible, desolate loneliness came to her. All love, life, desire was stripped from her being. She lay alone in her room, enduring the great crushing weight of that desert wind. It bore down upon her relentlessly, and she was raw inside with the beating of it upon her heart. Everything was gone now, yes, even the hate, the rebellion, the indomitable pride and will that had marred her life.

Finally, after many, many hours, something began in the wilderness that was her soul. It was like a sweet, clear call in the hollowness of a bitter eternity. It was very faint at first, but it seemed to promise relief from the unendurable anger of the sun and the wind; it was a slight hope of cool waters and soft, sweet meadows. It moved her just a little. She slid off the edge of the bed and onto her knees.

What would she do there? She had no words. She had no prayer. Indeed, she had no God to address. But she was on her knees, brought there by an incredible promise of life after a living death. She bowed her head. It was as though a cool stream—long dried up—sprang back limpid and sparkling, it's freshness defying the awful desolation of her desert. It began as a trickle, meandering its way through her spirit, touching and cooling the hurt places there. Where had it come from? Certainly she had no such stream of healing within her own wracked spirit.

It was as though her soul drank deeply and thirstily from it, and even as she wondered at it's source, the understanding was given her like a flash. It was completely a gift from God—a free gift, one she had not earned, in no way deserved, but a renewing balm from God. All her life, she had tried to be her own master, her own savior. Pride had prevented her from receiving help or counsel from anyone. Rebellion had dictated only wrong decisions, and her stubborn will had prevented her from correcting those decisions.

But in this instant, on her knees for the first time in many years, she saw it all so clearly. She could not be her own savior! Her resources were all gone. Meaning was stripped from her life. Sorrow and suffering, of such a magnitude that only a passionate soul as hers could comprehend, had shattered her heart. And from such poverty of soul she could not save herself. She would be forever at the mercy of the sun, the wind, the barrenness, the ravages of pain, but for the sweet, splendid gift from God that lifted her now from her desert of grief.

All the horror of these years with Jack, all the grief at losing her baby, the missed love with her parents; they were not God's fault as she had, in her pride, accused. She had brought them all on herself. If she had had the priesthood in her home to bless him, her John Paul might have lived as his uncle had. But she had lived in rebellion all her life, refusing to bend her knee, refusing to acknowledge any power above herself. Now it had brought her to this, the end of the last person she most loved. The realization was swift and uncompromising, yet brought with it instead of pain, a beautiful, uplifting illumination of her mind. It grew and grew until truth and knowledge enlightened her whole being, and she realized that truth can indeed make you free, for it enables you to see ultimate reality. With that kind of knowledge, you can be transformed from a servant of your faults and passions to the commander of your life for happiness. This truth, this unshackling of her soul, this invaluable fountainhead of freedom had come to her when she had no ability of her own left, no power to act, to speak, to do anything at all for herself. It had come from the very being she had so cursed and blamed. It had come, freely, lovingly, as the grace of God.

The words came out a whisper, not of uncertainty, but of sacredness, "Dear God, oh dear, dear Father of my soul. It is I who have been so wrong, so at fault. It is I! I know that now. Can you ever forgive me? My whole life has been a sin. I never acknowledged you when I had the chance, because I was too proud and too rebellious. How often you beckoned to me, invited me to come, and how often I turned away. Now there is nothing left for me to turn to. Everything that I love has been taken from me, and it is all my fault! All my own fault and no one else's. Can you ever, in all of your eternities forgive my sin, my pride, my curses? I . . . I need you, Heavenly Father. I never wanted to say that to anyone, but I come to Thee humbly now. Oh, how I need you. Be Thou my Father, and I will be Thy loving child, and all that Thou asketh of me I will do."

Charlotte knelt beside her bed for hours. Most of that time there were no words formulated or spoken. But there was communication. Light and love and peace came to her through a heavenly conduit, filling that barren wilderness of her heart with cool waters overflowing. She received it to herself, gratefully filling up her parched, thirsty soul. Never in her thirty-four years had she felt of this kind of sweetness. Later, as she pondered the experience, she recalled faint words, "He leadeth me beside still waters"—at last, still waters! She had never understood what it was that had converted her whole family and enabled them to endure the persecutions. Now it was hers; and she could not bear to move from the bed, to leave the room, even to speak, for fear it would dissipate, and she would be left alone again. At last she fell asleep, her head still bent over her hands. It was a deep, peaceful, restful sleep.

They buried John Paul in a little meadow near the river. It gave her somewhere to ride each day. Jack would not go with her. He told her he had said his good-byes when he placed the boy's body in the coffin and nailed it shut. That was all right with Charlotte. She cherished those rides alone. It was amazing the insights that now came to her. Undreamed of knowledge, unimagined concepts, always accompanied with a sense of light and illumination. She rode Shannon unhurriedly, content to meander. The world was as new and beautiful to her as when she was a child. She began to whistle again—even to sing. She sang the little snatches she could remember of the songs her parents had sung. She remembered most of "Rock of Ages" and "Onward Christian Soldiers". But her favorite was the last few lines of her mother's favorite hymn, "Til we meet, til we meet, til we meet at Jesus' feet."

The light came more frequently than she would ever have dreamed, teaching her all the things she had barred for so long. She was open now and accepting as a child. She questioned nothing, she rebelled against nothing, she asked for nothing, yet everything was being given her.

All the glory of the earth around her sang in its full, rich, summer voice, "You have lived for all eternity, and you will live on for all the rest of forever." She had lost sight of the immortality of the soul, yes, even her own soul. The dark years had convinced her there is only earth life with its trials and tribulations. Now the Spirit filled her with a vision of herself as she had been ages before, unfettered by the weaknesses of mortality, pure, refined as a spirit child of the great God of the universe. And now she saw the vision of what she could someday be, glorious beyond description. What strength and hope it gave her, that vision—a strength beyond any she could have conjured up from the determined stubbornness of her own personality.

Joy! It had never been hers before, never. Only briefly had she known happiness. Only occasionally had she recognized moments of happiness with her sister, Annie, with John Patrick and later John Paul. The closest thing to real joy she had known was with her babies in her arms. Now she could see that the relationship of love, of unselfish love and trust and service, was as close to the grand, spiritual experience of joy as she had ever been. How overpowering to know she might have joy even yet! She could conquer the danger of her own pride and the danger of Jack's twisted mind to experience that promise the scriptures made! Joy—man was created to have joy! And just as clearly as that concept illuminated her mind, was the assurance that joy came through one channel and one only. That channel she had finally accepted. Her prayers to her Heavenly Father were perfectly humble, unassuming as a child's, and His peace and love filled her with the first real joys of her life.

One day, as she sat beside the small grave, pulling absently at glossy green blades of grass, the final understanding came. Jack had been necessary to her life, for he was the only one who could have so tested her, the only one who could have brought her so low that she finally was humbled to the very dust. For her rebelliousness was a tragic flaw that might have kept her from ever achieving her ultimate potential. She saw it clearly all at once, and with her face awash with tears, she spoke her thanks to God for all the suffering, all the pain, that had torn from her the rebellious stubbornness, and for the comprehension His Holy Spirit had granted her.

Thus it was, that Johnny found his green-eyed, hot-tempered Charlotte and baptized her in the little pool of water they worked together to dam up, just as they had so many years before in Nauvoo. Johnny had asked in Montrose what the gossip was about Mrs. Boughtman. It was all over town, she and Jack had had a terrible fight over Connie months before. She had knocked him out with a chair and run away. Jack had sent various work-hands out looking for her ever since but couldn't locate her. Then last week, the little boy, John Paul, had been kicked in the head by a mule, and now Matthew was gone to try and find her. Boughtman's hired-hands loved to talk about the Boughtmans, and the town loved to hear it.

Johnny decided to stay around for a few days. He knew that Charlotte would come back for her boy if she were located. Sure enough, word came into town with the hired man one night, that Mrs. Boughtman was back and crazy with grief. The boy had died in her arms.

Johnny walked out to Boughtman's and hid himself in the bushes around the home, watching the windows for any glimpse of his sister. He would have gone to the door and confronted Jack, but he was afraid of what a violent scene might do to Charlotte. He kept hoping to see Jack ride off so he could get in to see her alone. Instead he saw Jack building a small coffin. Johnny stayed in the woods all night, watching Charlotte bedroom window for sight of her. Next morning, just after daylight, he was startled out of his hiding place by old Emery, now a permanent fixture on the Boughtman ranch. He didn't do much anymore, just boss the younger fellows around, but he was the only family Jack and Charlotte had. Emery nudged John in the ribs, then recognized him when he turned over, and apologized profusely.

"You better not stay here, Mr. John. If Boughtman should find you, no telling what he would do. He's just about as crazed as she is. Never seen a man like him carry on so over his younguns. Might think he didn't have no heart at all, 'til you see him with his children. Plum nuts over them. You go on back to town. When he leaves or she comes out to where you can see her, I'll send word."

Emery was as good as his word. A few days later he rode into town early one morning and said Mrs. Boughtman was going riding down by the river. Charlotte was lying face down on the cool earth, running her hand through the spring water when John Patrick walked quietly up behind her. She was deeply lost in thought until she saw his red head and freckled face smiling at her through the ripples in the stream.

"Johnny, Oh Johnny!" she screamed and jumped up to throw her arms about his neck.

"How did you know?" she searched his face for the answer.

"Didn't know 'til a few days ago."

"What are you doing here?" Her hands were on his face, his hair, as though she couldn't believe her eyes.

"I've been sent on another mission for the Church. I came out here to baptize you." He was grinning at her.

Her smile disappeared. Her eyes were sober and hauntingly sad. "Here I am. I'm ready at long last."

She was brand new, pure and clean, clear and shining as a star-splendid night. She told Johnny all that had happened in nine years. She spared herself nothing. The humiliation, the blasphemy, she told it all and wept at her own blindness. But through the tears, was a happiness born of hope and right and certainty that she was now doing what her Heavenly Father wanted her to do. She had only one wish, and she was almost reluctant to express it. She knew she had done nothing at all to deserve any special favors. Still, she voiced it to him after some urging. She wanted only to have another child so that she could raise it in the Gospel of Jesus Christ.

"Charlotte," he told her, holding her hand between his own. "The Lord will grant wishes made in righteousness, and I cannot think of anything more righteous than that."

Her face begged him for confirmation. "Do you really think He will trust me with another one? I've failed with all three I have already had."

"Yes, but this time will be different, won't it?"

"Oh yes," she assured him passionately. "It will be so different."

"Then have faith and God will grant it. Are you determined to stay with Jack, then, after all that has happened? I'm afraid for you, Char. Come, go with me. I'll take you out to Utah myself with one of the wagon trains."

She considered that a long time. Finally she spoke, "I'd love to go with you anywhere, Johnny, especially to live in the lap of the church. But I think my race is not yet run. Something tells me that I must stay, not only that I might have a baby, but its something about Jack. I tried to leave him once. I didn't want to come back but something pulls me back. Heaven knows I don't love him, it isn't that. But somehow I feel his life, his eternal life, for good or for ill depends on me. He must be allowed to come back, as I have done, or to finish his course, and I am mixed up with all that." She put her hand on his arm. "Don't be afraid for me, Johnny. Nothing can hurt me now, not really hurt me. Knowing God puts you beyond that."

"Now then," she said brightly, "let's talk about happier things. The family, tell me all about them. How is mother and Annie?"

They stayed until twilight approached, and he caught her up on all the years since they had been together. Margaret had her own little house. Her days were filled with visiting the sick, the widowed, baking and taking to neighbors. She had a school for young girls twice a week and taught them the inevitable needlepoint, crocheting, knitting, tatting, pottery making, and dress tailoring. She had had several offers of marriage, but would not accept any. "Patrick O'Neill was a man without successor," she always said.

"Now, Annie," Charlotte said anxiously.

"Annie," he said thoughtfully, "has become our territorial doctor. She married Jedediah Stringham. She was his third wife and found her place quickly in his family. It is hard for anyone not to love Annie. She had four children, and all but one of them died within a week. Finally, Brother Pratt told her she must not have any more or it would probably kill her, too. She protested until Jedediah told her he couldn't stand to risk losing her or to see another baby die. So she has turned doctor. She went away for two years to school in Denver, and now she rides miles and miles each week caring for the sick and expectant. She is best with children. Just her hands alone seem to cool a fever and soothe aching muscles. She sits with the sick ones by the hour, singing them songs, telling them the stories that we used to tell when we were kids."

"Johnny, tell Annie how much I love her, will you? And tell her I am going to be all right now. When times were bad, I used to cling to the memories of her and of you and Mama and Papa. There were lots of days when your shadows were as real to me as you are right now. Without those good memories, I couldn't have lived through it all. Funny, even though I clung to the dear memories, I couldn't see that it was the Church that made those people good. I understand now."

John Patrick stayed in Montrose for a week and saw his sister every day. Most of their hours were spent poring over the scriptures. Charlotte had much to learn and she was an apt pupil. John gave particular emphasis to the Book of Mormon and the Book of Commandments, a compiled record of Joseph's revelations. It was all elixir to Charlotte, the food of the gods for a hungry spirit. They never studied at home. Once they had been discussing a scripture in the Bible when Ruby entered the room. She had sat down and listened to their conversation a while, then left without a word. That night she told Jack that her Uncle Johnny was preaching to Mama. Jack brooded about it and questioned her stringently. She didn't tell him that she had already been baptized. Jack never stayed around when John came to visit. He was as polite and civil as his jealousy would permit, but he always left immediately on 'business'.

John sought out Matthew and tried to find a way to influence him for the Church, but after a few conversations, he knew that the boy had been poisoned too long by his father. Finally, as Hector reminded him, it was time to get on. They had been sent to Canada, not Montrose. (Johnny didn't really believe that. He knew in his heart that he had been sent to Charlotte at that particular time.) Reluctantly he said good-bye again to his sister. But this was a happier time than the last. She would be all right now, he knew. He left her with his blessing, and with his red hair blazing in the sun, he rode away one last time.

CHAPTER 11

Ten months later, Charlotte labored for four hours and brought forth a baby girl. She named her Annie. Jack made no objection. With each of their children he had had real enjoyment, but for this little Annie he felt nothing. She could not replace John Paul.

Annie was a doll of a baby—red, curly ringlets covering her head, deep blue eyes filled with wonder, as only a baby's eyes can be. She was completely absorbed in the sight of her mother's face, and Charlotte wrapped herself up in the joy of her gift from God.

Ruby was growing up fast and was anxious to be called a woman. She was envious of Matt, his independence and status in the adult world. He had become one of the best ranch hands on the place, and it was clear that Jack was grooming him to be the next foreman. The boy had a decisive mind, able to sum up a situation quickly and make a decision. Like his father, he almost never backed down. Also, like his father, he believed that animals need a heavy hand and a master not reluctant to use a whip.

Ruby took a great interest in the horses. Of course, there was not much else to occupy her. She had begun to spend more time in town of late. She frequently rode into town with Jack or Matt and was making friends there. Charlotte asked often about those friends, wishing for more intimacy with her daughter, but Ruby was non-committal.

"They are just girls, Hattie Williams, Deborah Mills, Alicia White. They're my age and we just sit in the swings in their yards and talk."

"Why don't you have them out here, Ruby? We could give a little party and they could stay overnight with you. Now that Emery and Lily are in their own house, and Matt is in the bunk house, we have plenty of room. Wouldn't that be fun?"

"Not really, Mama. Besides I don't think they'd come."

"Why not?"

"Well, you know the Boughtman family does have kind of a wild reputation." Ruby looked at her mother out of the corner of her eye. She knew that would affect her, and she was right. Charlotte's face took on a pinched, set look. It was the last time she suggested a party.

That was just what the girl wanted, because she neglected to tell her mother that each of those girls had brothers to whom she was more inclined than to their sisters. Her friendship with Alicia White was very shallow, but her friendship with her brother, Josh, was more substantial. He was about her height, for she had shot up in the last two years. There was only one young man in town who was taller than she. Josh White, Marshal Mills, and Sam Williams were best friends, so it was natural that they share the same girl. After one of her visits into town, they would spend the rest of the day whispering about Ruby Boughtman and her kisses.

Ruby was no fool. She knew what her drawing card was. Her figure was slim and willowy, her hair a black cloud to match her eyes, and her skin glowed with a fresh bloom of pink in her cheeks. She often stood in front of her mirror turning this way and that. She knew she would keep the interest of her three beaus only so long as she only permitted them a taste of what she had to offer, and, at fifteen, Ruby Boughtman was a scrumptious morsel.

Charlotte was concerned about Ruby and talked at length to Jack about her. He would not have the girl criticized and never failed to defend her. His favorite comeback was, "She's no wilder than you were at her age." She had no real evidence, but somehow Charlotte knew the wildness was different. It ate at her to see Ruby teasing the men on the ranch, wrapping her arms around their waists, sidling up as close to them as she could get.

Charlotte said one day, "Being so young, my dear, you probably don't realize the effect you have on the men when you stand so close, or wear your dresses cut so low."

"Oh really!" Ruby replied, her eyes wise, eyebrows up, the picture of surprise.

Charlotte glanced at her sharply, "Yes, really. Those men have work to do, and you have no business distracting them."

Ruby flipped back, "They don't seem to mind."

"But I do. I won't have my daughter labeled as a flirt and a tease."

"Like mother, like daughter, I guess."

Charlotte almost slapped her before she caught herself. "What exactly did you mean by that?" she asked tightly.

"Oh mother, you don't have to be so prim with me. I know all about your boyfriends. Don't you think we all know how you've shamed Papa all these years!"

Charlotte was seething and held onto her temper only by the sheerest exertion of will power. "Ye are a fool, little girl. I am to be censured in many things, but that is not one of them. And ye would do well not to place yereself where ye could be censured either."

She prayed that afternoon about it before talking to Jack. He was uncomfortable hearing his own fabrications coming back to him.

"Where did she get that idea?" Charlotte asked calmly.

He rubbed his hand over his jaw, "Well, uh, I don't know, Charly. Must of just said it to get your goat."

She stared into his eyes and saw a veil. She slowly began to suspect that he had told her children lies about her for a long time. She decided to let it go. "Talk to her, will you, Jack. You're the one she listens to."

"Sure, honey, be glad to. I never could see it before, but I guess you're right, our little girl is growing up. I'll talk to her."

He did!

"Ruby, honey, you must be careful what you say to your mother. She's mighty touchy on her past mistakes. I only told you about them because I wanted you to learn from them. That kind of thing can bring grief to everyone you love. You understand what I mean?" He looked at her meaningfully.

"Sure, Papa. I think I know what you mean. I'll be better, really I will. I wouldn't ever want to hurt you like she has."

"That's my baby. You know you are getting to be a mighty pretty young lady. I'll be losing you one day soon to some younger man, I guess."

"Oh no! You'll never lose me. I'll never love anybody more than you."

"Well, now, honey, someday you will. And that's how it's supposed to be."

She flashed her dark eyes at him coyly. "Do you really think I'm pretty?"

"Sure do," he said, patting her long, dark hair. "Just about steal my heart if you weren't my own daughter."

That night before she went to sleep, Ruby didn't think about Josh White or any of the boys in town. Instead, she made up dreams about some wonderful, tall, dark young man. He would come and steal her away, and he would be just like her own handsome Papa.

Until little Annie was three, Charlotte and Jack continued to have a fairly peaceful marriage. Charlotte had almost decided she had been impressed to stay because she would eventually influence Jack for good. She would never be blissfully happy with him, but they were trying hard to be patient with one another, and at least their relationship was no longer marked by violence. She had not seen his black side for a long time. But the pattern of the past was too strong to be entirely broken.

Their relatively peaceful world all started falling apart again one snowy Christmas when Jack had had too much to drink. Little Annie hung on her Mama's knee and whispered in her ear, "How come Daddy drinks? Heavenly Father doesn't like that, does he?"

"What're ya whisperin' about?" Jack roared. "Nobody whispers in this family. Everybody says what he thinks."

"She's just a baby," Charlotte said.

"She can talk. Well, let her talk out loud instead of whispering." He became indulgent. "Come here, little girl. Come talk to your Papa. Tell me some pretty secrets too, huh."

Annie looked pleadingly at her mother and still clung to her knees.

Jack spoke more sharply, "Come over here, missy."

She reluctantly obeyed, her mother nodding her head encouragingly.

"Now then," he said, coaxing her confidence. "Tell me what all the whispers were about."

The child had no ability to lie or to make up another answer. Her three-year-old-mind knew only to do what she was told. She sat on her father's knee, uncomfortable with his unexplored temper and strength. Finally she said in her tiniest voice, "Heavenly Father doesn't like you to drink."

He looked up, a blank surprise pasted on his face. No one said anything. He tried to understand and looked back at the little girl with red-copper ringlets, "What are you talking about?"

She was misled by the calmness of his voice. "Well, I asked Mama how come you drink that nasty whiskey when Heavenly Father tells us not to." She looked seriously into his face, "We should always do what He tells us to if we love Him."

He brushed her off his lap, and she went scurrying back to Charlotte. Jack towered over the mother and daughter. "Charlotte, what is all this talk about commandments and God? Don't tell me that holy, holy brother of yours poisoned your mind with this so-called gospel."

Charlotte said nothing, trying to discern what she should say or do.

Jack's voice became menacing, "My house won't have any damned Mormons living in it. Ever since the boy died, you've been acting different. I thought it was the grief that changed you. That had better be all. I never did hold with Mormons, and I don't hold with them now. I never will, so don't think it. Now, I can drink whiskey in my own house on Christmas Eve if I've a mind, and I don't want any squirt kid telling me I can't."

He looked around the room. Matt was standing at the window, looking over the snow-blanketed night. He was as tall as his father now, and unusually quiet. Ruby was seated in the high-backed satin chair, dressed in a deep gold gown Jack had given her. The skirt of it was spread out over the chair, and she sat stiffly, showing it off to the best advantage.

An idea came to him, "Come here, Matt, Ruby. It's time you learned to hold your liquor."

Charlotte stood up. "No," she said softly. "They're too young."

"Be damned if they are!" he bellowed. I took my first drink when I was thirteen. Matt ain't too young by a long shot." He put his arm around Ruby and squeezed her. "This princess ain't neither. A drop or two of whiskey'll put some pink in her cheeks, and fire in her blood."

"As if she needs any more of that," Charlotte said sardonically.

"You could use a little too, woman. You're getting positively dull, lately. Can't even get a rise out of you anymore. A man needs a good fight every now and then to keep his interest up. Come on, Charly, where's my bobcat? Let your hair down a little."

Ruby was watching her. Matt was not. She tried to catch his eye, but he wouldn't look at her. "No thanks. I'm going to bed. Come, Annie, dear."

She started to walk out of the room, Annie's trembling hand in hers. She called back over her shoulder, "Come along Ruby."

She called out impishly, "I'm not tired yet, Mama. Anyway, Papa's gotta have someone to keep him company."

Jack roared out good-naturedly, "Now there's a girl that knows how to make a man happy." Charlotte didn't look back.

In Annie's room, Charlotte dressed her little girl for bed.

"Mama," Annie asked, "Did I say the wrong thing?"

"No, sweetheart. You were right. If we love Him, we do whatever Heavenly Father tells us."

"Well, Papa doesn't. Does he love Heavenly Father?"

Charlotte considered a moment. "Well, he doesn't know Him very well."

"Why not? Can't you tell him like you tell me?"

"Daddy's don't like to be told, honey. You have to be very careful about teaching them. Example is the best way. So you just be a sweet girl and a good example to your Papa, and maybe he'll come around some day."

She hoped that was true. Jack kept the older two children up until after midnight, then the door slammed and she heard him riding off toward town. She came out of her room and stopped Ruby on her way to bed. The girl was light-headed.

"Where's your father going?"

"I suppose to town for a little drinking company . . . or something." She laughed and winked at her mother.

Charlotte closed her door and stood looking out the window with clenched hands until the anger had passed. Kneeling at her bed, she prayed and prayed for strength until, at last, a quiet, gentle peace swept over her, erasing the anger, replacing it with sadness. When Jack came home toward dawn, she never questioned him. She simply opened up the covers for him. He mumbled an apology for spoiling her Christmas, kissed her hair and cheeks, snuggled into her warmth, and promptly went to sleep.

After that, his visits to town took place more at night than in the day. Charlotte put them out of her mind. She knew after twenty years of marriage to him that he would never be satisfied with just one woman. Sometimes, she grew bitter for a moment over her young girlish dreams of being all in all to a man. Joseph's words had indeed proven prophetic. She would have much preferred knowing he had another wife.

As he drifted back to his old patterns, Jack began finding ways to humiliate her. His favorite was to insult her in front of the children. Any tenderness that had made its way into their marriage after John Paul died was rooted out, and they went back to the pattern of conqueror and vanquished. Jack seemed to thrive on it. But this time, one thing was different. She had always fought him, resisting to the last breath all the humiliation and assaults on her spirit. Now when he attacked, he found his opponent vanished, and there was no one to fight with. She simply withdrew. She would not battle; she would not argue. She refused to satisfy his need to master by simply not engaging in a contest. It was infuriating.

In fact, it was the most exasperating thing she had ever done to him. As he drove in his shafts only to find empty air, he grew more and more frustrated. Charlotte had always provided him the stimulus that kept him going, challenge! Now she would neither respond to his challenges nor provide him one. Their relationship deteriorated, and eventually he began swearing at her. The worse he behaved, the more guilty he felt and, consequently, the worse his behavior became. It was a familiar vicious circle that Charlotte knew well. She began to watch for the thunderstorm warnings.

By the following April, he had removed most of her clothes from the closet and left her only the oldest, shabbiest dresses. Then Jack hit upon another idea. In July, he let the Negro servant, Sophie, go. Sophie had kept the house, cooked the meals, and spread the juicy gossip about the Boughtmans all over town. Now, with her gone, Charlotte had to do her work and was tied to the house in a way she never had been before. The daily rides with Annie and Shannon came to a halt.

Annie grew impatient staying around the house, but Charlotte taught the little girl some things she could do to help. Soon Annie was dusting, sweeping the floors (after a fashion), setting the table, turning over the chicken while it fried, and scores of other little jobs that were now her mother's responsibility. While they worked, Charlotte would talk with her about the Church and the things she had learned in the last few years. She was determined to fulfill her promise to God, and she taught her daughter all that she could about the gospel of Jesus Christ. She told her personal experiences with the authorities of the Church, Joseph Smith in particular. She filled her receptive mind with stories about the angel who visited Joseph and the golden plates he translated to get the Book of Mormon. She told her about her Uncle John Patrick and the people he converted to the Church. Charlotte told Annie Bible stories and found she had a gift for making Jesus's love very real and personal. Jack never knew all that she was teaching their daughter. She cautioned Annie not to mention it to Papa. The girl was not tempted. She wanted no more to do with Jack than necessary. She was plainly afraid of him.

The storm warnings grew gradually bolder. Charlotte remained calm. Jack never failed to berate her for the slightest things. From time to time she left the chores, even knowing he would be angry, and took Annie walking in the woods. She watched the little girl grow and loved her fiercely. She was almost four now—a tender, beautiful child. Her hair was as red as old Patrick's had been, and her eyes, a soft brown like her namesakes'. Annie had a tendency to instant anger over childishly important things, and Charlotte treated that budding temper with the utmost delicacy and concern, knowing the grief it could bring. Annie often got angry with her father for yelling at her mother. Then Charlotte would quietly talk to her daughter.

"You know, honey, sometimes God lets us be tested in different ways just to see our strength. Like, often I give you jobs that seem very hard to you. Well, that's pushing you a little, but it encourages you to grow and be a little better than you normally would be."

"Is that what Papa is doing."

"In a way, yes. So I just try to meet the test and not get mad."

"Don't you ever get mad, Mama?" the little girl asked enviously.

"Oh, my goodness, yes. You have no idea how mad I have gotten at times, but it just hurts me in the end. So now I try very hard to stay calm."

Example was the best teacher and Annie wanted to be just like her mother.

It was the first of May. The winter had been long. Being cooped up with Jack around the house so much had been a constant drain on her spirit. He seemed to cast about in his mind for ways to get her to fight, or for ways to defeat her. She was tremendously relieved when the weather grew warmer. Spring had finally convinced the earth to rejoice, and Charlotte and Annie had been outside all morning planting flowers and clearing away debris that collected around the porch. The rest of the day she had spent making Annie a dress, cleaning house and cooking dinner, but when it was time to eat, Jack didn't come home from town with Ruby. They had gone on business and intended to be back by dinner time. It was now nine o'clock and they had not yet come in. She set their dinners in the cold oven and went to bed. Her head was filled with plans for two more dresses for Annie. The girl was sprouting fast. Her old dresses, that had fit just three months ago, were way up her legs now.

They had said prayers together as was their custom, and Charlotte had gone to bed as soon as Annie was tucked in. Charlotte had no trouble going to sleep, pausing for a few moments just before drifting off, to remember the way the world had looked in May when she was three years old.

In her dreams she was laughing. She and little Annie were wading in the stream and underneath the water were little fish that were nibbling at their toes. They were giggling. Their laughter became louder and louder until Charlotte awoke. Someone was laughing. In fact, there were several voices. It sounded like a party downstairs. She guessed that Jack must have brought back some friends from town. So she hurried and dressed, deciding to be a good hostess and greet them.

At the bottom of the stairs, she could see into the parlor. Jack had brought five or six young fellows with him. They were all rather obviously tipsy. Ruby was playing hostess, sloshing whiskey into their glasses, then sharing their drinks with robust high spirits. She played no favorites. There were five strapping, handsome young boys, and she intended to give them all a good time. She was dressed in a most revealing gown, cut low at the neck, lower than Charlotte had ever seen on her daughter. Ruby whirled, dipped, leaned, laughed, and scattered kisses like petals from a daisy. Even her father got his share. She filled his glass and put her arms about his neck, slurping from the opposite rim of the glass when he took his first taste.

Charlotte stood silently watching. Ruby was like a moon-mad gypsy, and her mother had the feeling she was really seeing her for the first time. She saw her through distant eyes. The girl was no daughter of hers. Old Patrick O'Neill flashed through her mind for a moment and the rage with which he would explode. But rage was not part of her then; there was, instead, a sick, defeated feeling inside. She made her move when Ruby draped herself around a young soldier, leaning over him and kissing him fully on the mouth.

"Jack, I see you decided to bring your party home."

"Oh, Charlotte, glad you came down! I was about to go up and get you so the boys could see what a woman I have for a wife. These here are some fun-lovin' fellas we met in town. That one young buck with Ruby on his lap is a soldier in the Union army of Mr. Lincoln's. Now ain't that proud! These here," he said, trying to encompass the other four within his long arms. "These here are some friends of Ruby's and mine from town. Marshal Mills, Josh White, Sam Williams—you know Sam's daddy, don't you honey?—and this is the young Allen boy. What's your name, son?"

The boy mumbled something incoherent.

They were all looking at her. Ruby insolently remained on her soldier boy's lap. He tried to wipe her kiss away and look half-way presentable, but she kept caressing his neck and hair with her long fingers.

"Well, I'm sure you're all welcome here, though the hour is getting on," Charlotte said as mildly as she could.

"Hell, it ain't late, yet! Ruby was gonna do us a dance. She's a fine little dancer, Mama. I never knowed it 'til tonight. You'd be right proud of her."

"I'd be prouder if she acted like a lady," Charlotte said, staring into her daughter's haughty eyes.

"What for?" the girl asked. "That isn't any fun. Never have no fun like that, would you soldier?" He was noticeably uncomfortable.

"Ruby, I think you've had a little too much to drink. You're forgetting yourself," Charlotte spoke more pointedly now.

The girl turned on her mother viciously, "Oh no I'm not. I'm finding myself. I can see who gets along in this world, and you, Mama dear, aren't one. Look at yourself in your old, ugly dresses, while Papa buys me and Connie anything we want."

Charlotte held tightly onto her shred of a temper. "Ruby, we'll discuss it tomorrow. You're not ready to join the men's world yet."

"I've been ready a long time, only you've been trying to keep me a baby. Don't do this, don't do that. Don't wear low cut dresses. It will disturb the men. Well, disturb and be damned. I like to disturb them, and they like it too. So, what are you gonna do?"

Charlotte looked over at Jack, thinking he must surely come to his senses and stop his daughter from acting like a tramp. He was watching her coldly, waiting to see what she would do. In an instant, she knew he had planned it. He was not drunk. He had led Ruby on, using his approval as bait, flattering her ego, striking at Charlotte with one of the few things she cared about, her children.

Her anger leapt upon her and in a flash she had the girl's arm in her hand, powerful from years of holding reins, and had the arm twisted up Ruby's back. She said between clenched teeth, "We'll go to your room now."

Ruby screamed at her, "And who's gonna make me stay?"

"Me. I'll tie you to the bedpost if I have to!"

There was a shout of rage from Ruby and a hoarse laugh from Jack. Charlotte stopped, realizing what she had said, and remembering with a sick heart when her father had said it to her. She wished he had done it. That pause was enough. Ruby jerked away, grabbed the soldier's hand, and pulled him after her to the door. Charlotte started after them, calling her to come back.

The girl paused only long enough to fling back at her, "I used to be jealous of you, Mama. Now I just pity you. I'm going to live my life the way I want, and I hope I never see you again. Come on soldier, what'd you say your name was? You sure are big, and handsome. Bye, Papa, come see me in town," she shouted, as they started off down the lane, she astride the soldier's horse, holding onto him from behind.

Charlotte sank down into the chair, defeated. Her heart was sick as she remembered her baby of seventeen years ago, a beautiful, dark-haired little girl that she had nursed and tickled and walked with in the tall grass. Tears slid unnoticed down her cheeks. She turned around. Jack was still standing by the fireplace watching her with amusement. The other boys were shuffling about, embarrassed, anxious to leave.

One by one they filed by a silent Mrs. Boughtman. Then there were just the two of them alone.

"You are a monster," she said incredulously to him. "You egged her on. You gave her your approval. You want to turn your own daughter into one of the town whores? How could you hate me that much? How could you hate me enough to destroy your own daughter?"

He exploded with pent up fury then, "You've never been mine. You've always thought you were better, smarter, stronger than me. Well, you're not! No woman alive is stronger than I am, and if I have to destroy everything to prove it I will."

"It won't work, Jack. I'm not afraid of you any more."

"See what I mean," he roared at her. "Always stronger, always superior!" He began circling her. His voice got low and vicious. "And something else. You're a damned, stinking Mormon, aren't you? Well, aren't you?"

"Yes," she answered quietly.

"Ha, hah! I knew it! I'll bet that brother of yours baptized you, didn't he?"

"Yes."

"That was a mistake, Charly. There's nothing on earth I hate more than Mormons."

"Why, you're one yourself. You told me you were baptized."

He slapped her hard, "Shut up! I ain't no Mormon! They dunked me once, but I wasn't really baptized."

"Oh yes, you were. You're ashamed to admit it. You were baptized, and you knew all along the Church was true, and that's why you hate it. It's true, and you're too much like me, rebellious, too hard-headed to live anyone's commandments, even God's. We always hate the things that condemn us."

"Shut up, I said."

"Papa told me that the worst persecutors of the Church were apostates because once a man knew the truth and turned away, all the light went out of him, and only darkness remained. Only, you've progressed, Jack. You've progressed into darkness, and Satan himself will possess your soul."

Jack slapped her again, her head snapping backward, but as he stood over her ready to beat her into submission, something bound his hands to his side. She looked at him steadily and spoke, "Don't you ever hit me again, Jack Boughtman."

Charlotte bought a small gun the next day. One of the fellows on the ranch had it. She gave him twice what it was worth and made him promise not to tell Jack. She had told John Patrick that Jack must be allowed to finish his course, and something told her the end would be soon. Jack had progressed into darkness while she had progressed into light. The two could not exist together and she was determined the darkness would never envelope her again.

She wrote to Victoria Palmerory, with whom she had kept up a correspondence. While she waited for a reply she made herself and Annie ready to leave quickly. She packed a small valise with changes of clothing, a sizeable stack of money, some finger foods to nibble, and the gun. They would steal away one night and meet Victoria. From there, they would go to St. Louis. She would wait until Jack was sure to be away a couple of days, then she would go.

While she was waiting for her answer from Victoria, she had a visitor. Michael Gailbraith walked up the lane one fine morning, tipped his fur hat and said, "Charlotte O'Neill Boughtman, I can na' forget ye. Ye are as beautiful as iver ye were."

She was overjoyed to see him. They took little Annie and went picnicking in the meadow beyond the woods. He told her about meeting a big, handsome Irishman on his way to Council Bluffs and being baptized in the Missouri River.

"So we rode together back to the Great Salt Lake. He's a rare one that brother o' yours. Like me own brother he seems."

"Have you built that little house in the woods for anyone, Michael?"

"Nay. I told ye I have na' been able to forget ye."

She blushed and looked off across the meadow. Annie was picking yellow flowers and making a necklace with them. "Isn't she a beauty, Michael?"

"That she is, just like unto her mother. Charlotte, are ye happy now with this Boughtman?"

She wondered how much she should tell him. "No, I'm not happy. In fact, I have corresponded with Victoria and, as soon as I get her response, I'll be leaving here, I hope."

His eyebrows shot up. "Leaving? That's a big step for a woman to take. Where will ye go? What will ye do?"

I do have some abilities beside riding and training horses. I have a good education and a head for business. I'll try to find work in St. Louis and support my girl."

He held her hands. "Charlotte," he said urgently. "Let me take keer o' ye and the wee one. I'll love her like me own. I'll love ye like me own flesh. I want to marry ye, woman, and I'll marry ye under the new and everlasting covenant, for all eternity. Every woman has the right to a man who wants her forever, and who will protect her and keep her forever. Any man what mistreats a woman like Boughtman has done ye, does na' deserve her. He'll niver be yere husband in God's eyes, ye know that."

She was confused. She knew she could love him, but was it right to run off with another man? To run away for protection of your child was one thing. To go off to marry another man was something else. She thought of Annie. She wondered how it might look later to the girl when she grew up.

"I must consider, Michael. Anyway, we're not ready to go yet. I know you're a good man, and I could love you happily. But I must think of what is best for the girl, as well as myself. Don't be hurt. Please understand."

"I do, I guess. I'll wait for as long as ye say. Only I can hope ye're answer is in me favor."

She kissed him on the cheek and smiled her appreciation into his eyes. On the edge of the forest, Matthew Boughtman watched.

Jack was in a black mood. He had become obsessed with breaking Charlotte, but she was like a reed in the wind. She bent clear to the ground but she didn't break. Surely, there was a point where she would no longer spring back. For twenty years, they had repeated the cycle. He would begin to get her to acknowledge him as master, and he would turn his back for a moment, and she would bounce back up. This time he would break that stubborn, hard-headed woman if it killed him and her too. All he really wanted was for her to come to him meekly and say, "Jack, you're the master here." Oh, she had accepted all the work and abuse he had piled on her, but something in her manner said that she was not beaten, she was merely flowing with the stream.

He had been drinking most of the day, mulling the problem over in his mind. The longer he thought, the more he hated her, the more determined he was to deal her such a blow that she would give in at last. Ruby had been sacrificed to this determination. She was living with Connie and the other saloon girls now. He knew she was soon to leave and follow her soldier back to the lines. That had hurt Charlotte, all right, but not enough for Jack. He loved his daughter in his own twisted way and hated to see her go, but that didn't matter! He cast about in his mind. What else, what else? He was so obsessed by it that he couldn't even get any satisfaction with his whiskey. Early in the evening, the answer came to him. One thing she had loved, probably more than she had ever loved him, was that damned horse, Shannon! A stinking Irish name if there ever was one.

I'll take him and trade him down the river, he thought. Then another idea more insidious came. The plan began forming, and he worked it out sitting on the porch in front of the saloon. It was one o'clock in the morning when Jack rode quietly onto his ranch. All lights were out, and he hoped she was sound asleep. Just to make sure he crept upstairs and cracked the door. She was not there. He went down the hall and cracked the door to Annie's room. There were mother and daughter wrapped in each other's arms, sleeping soundly. Back in Charlotte's room, he hunted up the cloak she always wore, doused himself with her perfume and wrapped himself in the cloak.

The stable was black. There were three horses in their stalls. Shannon was lying on the hay-matted floor in the third stall. When the door opened, he raised his big head drowsily. Before the horse could get a look at him, Jack was behind him, tying a rag over his eyes, and securing it tightly beneath the jaws. He left the lantern dark while he led the other two horses outside and tied them up at the end of the corral. Then he went back in.

Shannon had struggled to his feet, nervous, wondering what was happening. He had smelled his mistress' perfume and her personal scent in the cloak. There was another scent, too, but the perfume covered it up. He felt gentle fingers on him and whinnied, thinking it was Charlotte. Then came the soft cotton rope she always used to lead him, and the brush she used to brush him. He was anxious. He wanted to see her. He twisted his head, jerking it, trying to work off the blindfold. She didn't speak to him, but again he felt the gentle fingers soothing, quieting, loving him. He calmed down and submitted to whatever she intended.

When the lash came down, poor Shannon was numbed. And it fell again and again. With the perfume in his nostrils and the jerking of the cotton rope over his nose he was thoroughly convinced it was Charlotte. He took several lashes with a pathetic whinny, but they grew worse. He reared up. The rope choked him down, and the whip bit again and again.

It took Jack better than half an hour to whip Shannon to his knees. The horse was now nearly twenty two years old. His age had taken its toll, and his strength was quickly dwindling. Jack was thorough and the heart of the big chestnut broke first. His whinnies were strangled as if choked by tears. Jack left him, a bloody heap with battered legs and deep slashes where the whip had bitten time after time. Shannon didn't move.

Matthew rode into town at dawn, when his work was done, and brought Jack two juicy tidbits of news. Someone had gotten into the stable and almost killed Shannon. He was beaten nearly to death. The other news—a man had visited his mother. They had taken Annie and gone picnicking in the meadow, and he had seen her kiss him.

Jack flew into a rage. He grabbed his son by the shoulders and shook him. "Are you sure? Who was he? Did she actually kiss him? Was there anything more? Did they do anything more?"

Matt was frightened. Yes, she had kissed him. Well, yes, it had been a long kiss . . . on the mouth. Yes, (he said so didn't he . . .) on the mouth. No, they hadn't made love. (He couldn't bring himself to say that.) He thought it might be that man who had found her in the woods the time when she was gone for so long. He thought the name was something like Gailbreth, or sounded like it anyway.

Jack ranted, he raved, he stomped, he broke up a chair and pounded the wall with it. Finally, he calmed down enough to ask Matthew if he had told Charlotte about Shannon.

No, Matt had been too heartsick to tell her. The horse was a pathetic sight, and he didn't want to be there when she saw him.

"Why spare her anything?" Jack yelled at his son, and Matt drew back, startled.

After a while, Jack ran out of questions and sat thinking. What to do? No wife of his was gonna make a fool of him. He had to decide what to do with her. He and Matt went to the hotel for some breakfast, but Jack hardly ate a thing. Matt could see something was brewing inside his father, and began to wish he hadn't told him about Gailbraith. He had seen his father in his black hole before, and it wasn't pretty. He couldn't understand why his mother persisted in tormenting his Pa.

It was about two o'clock when they went back over to the saloon. They sat drinking at the bar. Jack downed his whiskey's one after another. Matt nursed one along. He was beginning to get nervous about his father.

There were two men seated at a table in the corner. After awhile, Matt began looking more closely at the men. It was then he thought he recognized the blonde man in the corner. Jack noticed him staring at the table and said, "What are ya lookin' at?"

Hesitantly the boy said, "I think that's him."

"Who?"

"That yellow-headed fellow over there. I think that's the man that was with mother yesterday."

Jack spun around and stared. "Well, are you just guessing or are you sure?"

"I'm pretty sure. That other fellow was wearing a fur hat and was dressed up in them frontier clothes."

"Let's go see." Jack got up slowly, like a cougar eyeing his prey. He walked over to the table. Michael looked up, unconcerned.

"Yes. Are ye a wantin' something?" Michael spoke congenially.

"Is your name Gailbreth?"

"Gailbraith," he answered. "Who's wantin' to know?"

"Jack Boughtman, damn your hide. You the man that's been fooling around with my wife?"

Michael stared straight into Jack's deadly gaze. "Ye must be the bear that treats a woman like a dog. I niver met a mon who proved his monhood by beating a woman and humiliating her in front of his friends. The only animal what would do that would be a bastard cur. Is that ye, Boughtman?"

Jack didn't answer. His fist shot out and connected with Michael's temple. The trapper went sprawling. He came up fast and hard from the floor and caught Boughtman in the groin with his head. Jack moaned, his stomach sick from his toes to his head, but his fury was greater than his pain, and before Gailbraith could get to his feet, Jack kicked. It knocked the breath out of Michael and he reached for his midriff. As Jack bent over him to haul him to his feet, the years of animal wariness served the trapper well. His fist came from nowhere and sent blood spurting from Jack's nose. It was broken. Jack cursed, wiping the blood away. The barroom was deathly quiet, the men standing well back, watching the fight. While Jack was stunned for a moment, Michael spun around, tucking his shoulder, and wading into the bigger man's gut. He had the advantage for a moment, and had Jack gasping for breath, but the unearthly rage within Jack would not be stayed. A huge fist came down on Gailbraith's head, and he folded at the knees. It came again and Michael was out. Boughtman planted his great, booted foot in a full-force kick against his temple. It shattered the scull and drove a splinter into his brain.

When Matt saw the trapper out, he pulled his father back before he could pummel him to pieces. Outside in the bright afternoon sun, Jack was still in a fury, "I'm not finished with him, yet. But right now, I wanta see your mother."

Matt tried to get him to wait. He knew his father was worked up dangerously, but Jack would have none of it. "I'm going to the ranch. You can come if you want, or stay, but I'm going.

Charlotte heard them ride up. She and Annie were beginning to prepare dinner. Some instinct told her to get Annie out of the way. She sent her upstairs to her room.

Jack came crashing in the front door, Matthew trailing behind.

"Charlotte, who is this Gailbraith fellow you've been sneaking around with?"

She caught her breath. Glancing at Matt, she thought she understood the dismal look on his face. Measuring her words carefully she began, "He's a friend that saved my life once."

Jack sneered, his face an inch from hers, "And were you properly grateful?"

"No, not very much at first."

"Damn you, cheating woman, don't you think I know you've been seeing this wild man behind my back, kissing him and Lord knows what else. Matt here saw you." He grinned triumphantly now. "Your own son saw you kissing the man." He turned to Matt. "Guess that about proves what I've told you all along. She's always been like this, a sneak and a tramp, and now you've seen it for yourself."

Matt's face was abject misery. He knew he had lied and couldn't now back out of it. He dared not look into his mother's face.

Her voice was low, "Is that what you've told them all their lives? Is that how you've turned them against me? I should have known." Her eyes were flashing green now, and her voice was thick with Irish brogue. "I should have known ye would use anything, even lies to take my babies away from me. Now Ruby is gone, a camp follower, and Matt thinks I'm a common tramp. Oh, ye had yere way all right. Ye must be a miserable, miserable man to want everyone else as miserable as yere self. Ye are the devil's own son, Jack Boughtman, and he's welcome to ye."

She turned quickly and started up the stairs. Jack caught her arm, jerked her around and raised his hand to slap her. She spoke sharply, "I told you last time, don't you ever hit me again."

Matt remembered the rage he had just seen in his father, and a man lay unconscious, maybe dead, on the floor because of it. "Papa," he grasped Jack's uplifted arm with both hands. "Papa wait! Don't hit her. She's not worth getting angry over." He glanced into his mother's eyes, and she saw an apology there. "Come on, Pa, I want to show you the papers for a deal I just made."

Jack pulled away. "Don't wanta see no papers."

"Well then," Matt cast about desperately in his mind. "Unlock the money drawer. There are some horses I been thinking about buying." Matt pulled him into the parlor and Charlotte dashed upstairs.

The time had come! Annie was sitting on the edge of her bed. With all the faith of her little soul, she prayed that they could leave and never come back. She was frightened by all the shouting. Charlotte went straight to the closet and took out the little traveling case she had there.

She took out the pistol and wrapped it carefully in her shawl. Holding it in one hand and the traveling case in the other, she said, "Come on, sweetheart, let's go now. Don't be afraid, honey, just hold on to my skirt and hurry as fast as you can."

They went quickly, but quietly, down the stairs. She peered into the parlor, briefly. Matt still held Jack's attention. They were counting the money. His back was to her, and, in a second, she and Annie were out the door, sprinting across the yard to the stable.

She threw open the doors and rushed to Shannon's stall. Matt had given strict instructions to the ranch hands not to tell Charlotte about her horse. He had intended to clean Shannon up before she saw him. But he never had the chance. She stopped in the door of the stall and stared down at her old friend. He was lying down, blood caked, legs covered with huge knots, some of them bleeding, his big aristocratic head limp.

"Oh, no!" she cried in awe over the horror of it. "Oh, dear heaven above, who could do such a thing. Shannon, my love, my beauty, Shannon!" She moved closer to the stall, weeping for her long-time, best friend. The horse started, snorting in terror. He was weak, but still his legs started thrashing as he tried to get up. His eyes grew wild smelling that perfume and hearing her voice.

"What's wrong! What's wrong, old friend. It's just me. I'll help you. Quiet boy, quiet now, let me see how bad it is."

As she reached out to touch him, the horse went wild. His ears back, he was jerking and thrashing, snorting and striking out at her. She backed away, incredulous. She had never seen him afraid of her. She was crying unashamedly now. "Oh, Shannon, Shannon, my beauty. It's me! I won't hurt you. I'd never hurt you."

She heard footsteps at the doorway. Jack called out, "Oh yeah. You're the one who did it."

"What do you mean?" she asked, uncomprehending.

"Why do you think he's so afraid of you? He knows your scent, doesn't he? He thinks it was you who beat him." And Jack picked up her cloak he had tossed outside the stable door.

It took a second for it to hit her. "My perfume. The bottle was empty. I thought Annie had gotten into it. He does know my perfume. He thinks I did it. But it was you, wasn't it?"

She was keenly aware of the pistol concealed within her shawl. Rage swept her and she wanted to scream at him, to take her little pistol and put an end to his evil, but Annie was crying now, hanging onto her skirts.

The last possible second she pulled herself back from the brink of murderous hate.

She put her arm around the girl and swept past him, out onto the lane. Matt was almost to the stable. They passed him on the way. She spoke to him once more. "Matt, if you've any sense, get out of here. Get away from this madman."

Jack called from the doorway. "Where do you think you're going, ma'am? I didn't say you could go."

She didn't answer. She and the little girl were hurrying as fast as they could toward the big road. Jack yelled out, "You'll never get away from me, Charly. I'm your husband, remember?" He raised his hand with a whip in it and started after them. But before he could catch them, Matt spun him around and knocked him to the ground.

Jack was taken completely by surprise. He sat stunned on the ground.

"Leave her alone, Pa. Let her go if she wants to."

"No, by hell! She's not getting away with it. She belongs to me and she'll have to learn that. Get out of my way, boy."

He stood up and Matt hit him again. This time Jack was not taken off guard. He came back and the two of them wrestled in the dirt. Matt was young and strong from years of work on the ranch. He was also determined. He had begun to understand, at last, that his father had turned him from his mother, and he was afraid for her if Jack should catch her. But the older man was filled with a power beyond his age. Darkness had seized him again, and he cared not for his son, for anything, except to destroy Charlotte.

After several minutes of wrestling, his big fist caught Matt's jaw and knocked him cold onto the ground. Jack led a paint horse out of his stall and hitched him to the buggy. Before he climbed in, he grabbed his whip and tossed it in. He clucked to the horse and started out at a trot. He found them a little way down the road, half-running, half-walking, Charlotte pulling the little girl along.

She had no clear plan of what to do. She had thought of striking out through the woods, but she thought the road would be smooth and easier going. A little further on was a secret path she and Annie knew that would eventually take them into town. Now she saw how foolish it was to think she could ever get away on foot. All she could do was bluff it out and hope for the best. She heard the wagon and hoofbeats behind them. Annie started crying again.

"Shh," Charlotte said absently.

"Mama, I'm scared."

"It's all right, honey. I won't let him hurt you."

The only certainty she had of that was the little pistol she was now unwrapping. She wouldn't use it if she didn't have to."

He slowed the horses to a walk behind them and started taunting. "Charlotte! Oh, Charlotte O'Neill, the hot-blooded, hot-tempered, green-eyed Mormon 'lady'. Where're you going, Mormon lady? To your lover? I put him out cold in the saloon. He can't help you. Come on, Charly, my red-headed girl, where's your fighting blood? You used to be some kind of fun. I don't like this new, mousy lady."

She wouldn't say a word to him. She just kept hurrying Annie along.

"How'd you like your horse, little lady. Just like God, I give and I take away. He'll never be your horse again, will he? Damned right. He thinks it was you. You know I cursed you with every lash of that whip."

Still, she didn't reply, nor would she look back. She heard his voice growing ever more hateful and mean. Still, she was unprepared for the lash of the whip when it cracked against her shoulder blades. She gasped, and turned once. Jack was totally wild. Standing up in the buggy with the whip in his hand, long black hair waving in the breeze, he looked like Satan himself. The whip lashed out again, and Charlotte staggered. Annie screamed and reached for her mother. Charlotte kept her feet and hurried on. The whip spoke again and wrapped itself around her neck. It gagged her, and she had a moment of desperation when she thought she would black out or choke to death. She tore it off and picked up Annie and started to run.

Jack's laughter boomed out behind them, and he clucked to the horses. Drawing up alongside them, he snaked the whip out once more. This time it hit Annie, cutting through the sleeve of her dress and bringing blood to her little shoulder. Annie screamed in pain, and that was the end for Charlotte. She would take no more. Charlotte stopped, lowered her daughter to the ground and bent over her. Beneath the shawl she pulled back the hammer of the gun. The whip sang out again, cutting into her legs. So intent was she on what she was about to do, she hardly noticed the blood or the pain. "Father, forgive me if you can," she whispered. Then she raised the pistol, turned and shot him.

She thought for a moment she had missed. His face registered only surprise. Then a little spot of blood began spreading below his shoulder. He looked down at it, bewildered, then covered it with his hand. She knelt, in the dust of the road, with her arms around Annie. All of a sudden, he slumped over. She looked down at the little girl who was crying uncontrollably.

"You're all right, baby, you're all right," she said gently.

Slowly, she walked over to him, and stood looking down at this madman that had been her husband. After a moment, reason returned, and she knew they had to get far away. She stood waiting for him to fall out of the buggy, but he didn't. He was slumped over precariously, but still, the buggy was not free. Carefully, she approached his body. Was he dead? Would he suddenly leap on her? Then swiftly making up her mind, she gave him a mighty push, and he rolled off into the dust of the road. She ran and picked Annie up and flung her into the buggy. Clucking to the horse, they started off in a trot and soon a full gallop. She looked back once. Jack's body was a tiny heap in the dirt. She was free at long, long last!

CHAPTER 12

Summertime in St. Louis was a mess. The streets were continually stirred up with a multitude of horse's hooves, wagon wheels and people scurrying to and fro, getting supplies for the long trip west. Hundreds of wagon trains a year went out from St. Louis to the west, and they went between March and August. To start any later would put you in danger of being caught in the mountains with the first winter blizzards. So, through the most humid part of the year, with thunderstorms turning dust to mud, wagons were readied and business hummed.

Charlotte, Annie, and Victoria Palmerory lost themselves in the melee of the frontier outpost, St. Louis. Victoria's second cousin, Elberta Longstroth, owned a prosperous rooming house on the residential outskirts of the growing city. She made a place for the two women and little girl, directed Charlotte in her efforts to find work, and sat by the hour listening to Victoria's recitation of the horror her younger friend had endured.

They changed their names. Charlotte applied for jobs under the name of Susanna Peterson, and gave out the story that she was a relative of Elberta's. It took her only two weeks to find work. Her story was a pathetic one to most men with whom she applied. She told them she had a small, four-year-old girl, that her husband had died, and her parents had also died on the frontier many years before. She rarely left a place of business that the owners were not impressed with her abilities and sympathetic with her plight. Still, women were rarely hired. Given a choice, men were always hired first.

She struck gold one day when she went to the general store of Mr. Isaac Cutler. The good man was ill that day. He was abed on the small cot in the back of his store, and the doors of his business were closed. When the coughing fit took him, he would shake and choke and double up with the violence of it. He was a thin, little man with a bald head and a smart mustache. He would do no business today. He was more than annoyed when the rapping at his door stubbornly continued. Most people knocked for a couple of minutes and then gave it up and left. This person continued for ten minutes knocking, waiting, knocking again. At last, he stumbled to the door between coughs, jerked open the door, and began to lambast the impudent intruder. He got out only the first few of his well-chosen words before he really looked at Charlotte. The intruder was a slender, willow of a woman with reddish-gold hair that once must have been beautiful. Now the gold was being replaced by silver and the sheen of copper was dulled. Most disconcerting, were her eyes, calm as a country lake, gray with green flecks that made them sparkle and shine. He stumbled to a stop in his tirade, then asked more civilly what she wanted, forewarning her that he was not open for business that day.

"Why not?" she asked.

"Because I'm sick, my dear lady," he replied.

"Well then, who is going to look after your business? You'll lose money and customers."

He grumbled his assent to that, and explained it would have to be so.

"Oh no," she replied politely, squeezing past him into the store. Spotting an apron hanging on a nail on the wall, she slipped it over her head, tied it precisely in the back, and turned to a nonplused Isaac. She informed him that she was one of the best among shopkeepers. She had a way with customers, a keen mind for stock and detail, and an excellent head for figures. Her real talent was in keeping the books and ordering the supplies. Before he could protest, a man and woman came through the now-open door. Charlotte waved him back to his cot and took over the task of serving them. Between coughs, he listened and decided she was right; she did have a way with customers. The only thing she had forgotten to mention was her ingenuity in answering questions that she hadn't the first smattering of information about. At the end of the day, he listened to her story, and she had herself a job.

Victoria found her niche with her cousin. They complimented each other perfectly. Elberta was an elegant cook, her dishes were masterpieces, but her window sills, closets and corners were disasters. So, in return for their rooms, Victoria took over the work of housekeeping.

Charlotte paid for their food and any extras with her salary. Isaac paid her four dollars a week, really generous wages, especially for a woman. She was able to put away a dollar a week as a savings and preparation for Annie's trip. She told no one, not even Victoria, but she had already determined that by the time Annie was old enough, say fifteen, she would send her to John Patrick in Utah. In her traveling case, she had four hundred and fifty dollars. It would not be spent. It was Annie's security.

The months went by, and the great fear of being found, or of facing Jack again, died away. Charlotte concluded she must have killed him with her shot. The thought did not cause her grief, except as to whether God would forgive the murder. She spent weeks praying about it and finally gained peace. For Jack, she could feel no sorrow or pity, but, mercifully, there was also no hatred or bitterness left. She concentrated on making her daughter's life as happy as possible.

The first few months were like a dream. To have a cheerful home and bright, kind faces around her made the nightmare of her last few years with Jack slowly fade. There were times, seated around that little breakfast table in Elberta's frilly kitchen, that Charlotte's eyes would fill with tears. She would excuse herself quickly to attend to other matters. They were the tears of relief. It was almost too sweet, too good. She was afraid, for better than a year, to take a deep breath, whistle a favorite tune or swing hands with Annie as they walked. But it came. Slowly the knots of anxiety, guilt and fear let go. And one fine June morning Isaac Cutler said, "You're whistling! Didn't know women could whistle."

"Am I?" she asked, opening the little closet door and storing her bag inside? Smiling, she tucked away a strand of auburn hair and recalled, "My father taught me. He loved to whistle, nothing in particular really, just the song of his heart."

Isaac peered at her, "Well, it's about time I heard a song out of your heart. I was beginning to think you weren't happy here with me and my little store."

She cocked her head at him and smiled knowingly, "Now Mr. Cutler, you know very well that's not true. I fairly bloom when I walk in your door, and I know I'm not very good at hiding my feelings."

Old Cutler scratched at his tight little mustache, and smiled a wry smile at his young protege. "Well, anyway, I'm glad to hear you whistling, Susanna. It promises a wonderful day, don't you think?"

Cutler was a man generally content with the small, comfortable lot in life he had been accorded by a benevolent supreme being. However, in the five years since his wife's death, he had begun to grumble quite a little bit, only because he missed the many small attentions of that little lady. Now there was no one to tell him he was clever, or kind, or lovable, and he had taken to grumbling to himself when no one was around. Not ill-tempered, really, he simply wanted a few pats on the head like, an old domestic family terrier that whines and tries to get under his master's hand.

Charlotte was not put off for an instant. He could grumble at her one minute, and she would solemnly confess that he was quite right, and she was only half as competent as he with ledgers. Wouldn't he please be patient with her. The next minute, after he had searched the ledger and found no mistake at all, he would look into those grey-green eyes and mumble something about a mistake yesterday. Suddenly, those eyes would begin to twinkle, and she would confess the sin.

"Oh Isaac, you are perfectly right. I changed it to the right figures last night by candlelight when you weren't looking."

"Hrumph," he would mumble, knowing full well she had had no access to the ledgers after hours.

Soon the grumbling faded away, business improved, the little store was tidy and clean, and cheery smiles greeted anyone who opened the door. Isaac was a fair man and very soft hearted with favorites. After a year when his ledgers showed an increase of twenty-five percent, he called her back as she was bundled up and leaving for the day.

"Susanna, before you go, I want to give you your wages." He was nervous and the packet trembling as he held it out to her.

Her heart dropped, and she wondered wildly for a moment if he meant to let her go. But no, as she opened the packet, a quick glance told her that he had given her more than her usual wages. She looked at him in surprise.

"Just a little something extra. With Christmas coming on, I thought you might need a little extra. And besides," he added quickly, lest she ascribe too much to his good nature, "Besides you have earned it for me. My business has increased substantially since you bullied me into hiring you."

On impulse, Charlotte kissed his cheek in thanks. It pleased him enormously, and he made a great show of hemming and hawing, and patting her on the back as she left.

After a few years, Charlotte stopped worrying for fear her new life would dissolve beneath her feet. Annie grew like a sunflower, and her religious lessons never ceased. The real joy of living in St. Louis was not the air of excitement and importance, but the fact that groups of Mormon immigrants from England and the British Isles came up the river headed for Kanesville and finally to Zion. Charlotte would visit their camps and attend evening prayer and religious services on the Sabbath.

Cutler's store soon gained the reputation among those groups as the only fair place to trade. Whether or not that is true, is irrelevant. They simply felt more comfortable dealing with a member of the Church, as it was not uncommon to be charged double the price elsewhere if it were known they were Mormons.

Annie grew up a polite, spiritual girl who alternately exhibited great spurts of vivacity and somberness. At times, when she fell into a mood of depression, Charlotte would ask her what she was thinking about. It was usually that dreadful day of fear, horror over Shannon, pain, and seeing her father dead in the road. As she grew older, the quiet moods changed to deep reflections on right and wrong, purposes of life, and joy and pain. She was a girl who was old beyond her years, a deep well that contained ripples of her mother's sorrows and her own observations.

With Annie, Charlotte found the closeness she had only tasted with her sister. They shared everything, and as Annie reached fourteen, they even shared their clothes. Charlotte watched her daughter grow with a mixture of pride and reluctance. She knew the time must come when she would give her up. In Utah, Annie would meet and marry a fine man. John Patrick would see to that! Most of all, she would be permanently out of reach for Jack, if he still lived.

On Annie's fifteenth birthday, her mother showed her the small cache and told her what it was for. The girl was at first resistant to the idea of going away and leaving her mother.

"Can't you come, too?" she asked.

"Honey, I don't think I could make it."

"What do you mean. You're as strong as I am?"

"Well, not exactly. I'd like to think I'm still a young twenty-year-old girl, but a few things belie that. See these little crow's-feet around my eyes. They tell me the truth. I'm an old crow, not a young chickadee."

They laughed together, but Annie grew serious again. "Mother you never had much happiness in your life, did you?"

Charlotte started with surprise. "Why, darling, that's not true. It's not true at all. These last ten years with you and Victoria have been heaven on earth. I have thanked God every day for the gift of life. Just to be able to love you and to help you grow and to be so proud of you, who could ask for more happiness than that! And even before, when I lived on the ranch all those years, I had some happiness. Oh, there were some bad times. I know you remember a little of those, but there were lots of good times, too, dear. Your Papa wasn't always so wild. I think he had gone a little crazy, and perhaps is not to be judged, at least by us. No, there were lots of wonderful times, riding Shannon through the woods and meadows, picnicking by the stream, Christmas by the fireplace with all the world covered up with its blanket of white. Oh, and there were marvelous times when I was a girl with your Uncle Johnny and your Aunt Annie and the rest of the family. There was lots of love in my family, darling. Some of it I've tried to pass along to you."

She sat with her arm around her daughter. "I don't want you to ever think I haven't had my share of happiness. I have had more, really, than I deserved. Heavenly Father has been good to me."

Annie was quiet for a few minutes thinking about the trip. Then she confided, "I'll be scared to go alone. I'd rather stay here with you."

"Oh fiddle!" Charlotte exclaimed. "Oh, fiddle dee dee! Scared? Of riding on that big train, with a nice Bishop and his wife to help you and watch out for you? You'll get to hear the clatter of the wheels and see herds of buffalo that we've been reading about. You don't even have to walk like the Saints used to. You'll get to whiz along some iron rails, sleep when you want to, and stretch your limbs when the train stops. If you stayed here with me, you'd never get to meet our family. I want you to know the girl you're named after. She's a doctor, you know, and travels all over the Territory."

She reached for her daughter's hands and said intensely, "Oh Annie, take your chance! Go out into the world and do all the things I never got to do. See how the stars look a thousand miles away. See if the grass is just as thick and lush, and if the rabbits hop just as fast out in Utah. Write back to me and tell me how a mountain looks. Go to the conferences and listen to your prophet and send me back his words. And honey, if you ever doubt that this Church is heaven sent, or that your Heavenly Father lives and cares about us more than we do ourselves, just remember that your mother found it out the hardest way possible. Find a good man, a true man, and never mind if he has a dozen wives if he loves you, if he is gentle, and if you want to live your life with him."

Annie took that advice. She kissed her mother good-bye one April morning and started out by wagon to Omaha, Nebraska with a company of Latter-day Saints, where they would go by train to the great Salt Lake. She cried for most of the day until, she finally decided her mother wouldn't want that, and, thereafter, she seldom shed another tear over their separation, though her tender heart was often torn by longing. She wanted to be all that her mother wanted her to be. Annie probably knew her mother better than anyone else did, and she was convinced she was a queen among women.

Charlotte gave her gift back to God as she had long ago promised she would do. She had given her daughter the very best of what was in her and hoped it would be enough to see her along life's way. Only one thing she had not shared with Annie. She had known for a year she was not well. She had begun coughing as Isaac did. Just before Annie left, a doctor told her he thought she had the initial stages of lung disease. If she had had any subconscious thoughts of keeping Annie with her instead of sending her west, the doctor's report wiped them away.

She worked as long as she was able, but within two years, she was too weak and ill to go on. She accepted it all with patience and even anticipation. The only thing she asked of God was that she not outlive her remaining savings. She didn't want to be a burden on Victoria and Elberta.

There were many days when Charlotte was quite well and enjoyed sitting by the fire, or, in the spring, sitting outside with the flowers and willow tree. Life had been good these past few years, and she wished nothing more than she had had. Occasionally she thought of Michael and might have wished to have had a gentle man like him in her life. Still, he would have taken time away from Annie, and she was the most important one to Charlotte.

In the last year of her life, she had only one more heartache. Ruby came through St. Louis. She was with a soldier, as always, though not the same one she had run off with years before. She rode behind him on his horse, and she wasn't laughing anymore. Charlotte almost didn't recognize her, her pallor was bad, her hair was wild, and her eyes had a look of sickness. They were passing slowly by Elberta's rooming house one day when Charlotte was resting outside, rocking, crocheting, and humming. She could hardly believe her eyes, but it was Ruby, and her daughter saw her, too. The horse stopped and backed up, after Ruby whispered in the soldier's ear. Ruby slid down its side and stood shakily for a moment, regaining her balance. Then she picked her way to the porch. Charlotte could see she was in pain.

"Are you Charlotte Boughtman?" the girl whispered.

"Yes, Ruby. It's been a long time, dear. We've both changed. Are you ill?"

Ruby was shivering. "Yes. He's taking me to the doctor to see what's wrong." She mounted the steps. "Do you live here?"

"Yes, with some friends. Won't you come in and stay with us awhile?"

Ruby backed away, quickly, "No, no! I didn't come for your help. I just wondered if it was you." She picked her way laboriously back to the horse, and the soldier dragged her up behind him.

Charlotte stood up and started toward her. "Ruby, Ruby, dear, wait . . . don't go." But the horse had already started off, and Ruby didn't look back.

Later that night while Victoria, Elberta and Charlotte were enjoying a late evening snack, sitting beside a small fire, a loud thump and a moan was heard outside the front door. Elberta reached it first and opened it, finding a woman collapsed on the porch. A horse was just disappearing around the big oak tree down the street.

Charlotte cried out, "Ruby!", and hurried to raise her daughter up from the porch. The three women lifted her and carried her inside, where they laid her on the settee. She was feverish and shivering, and she had a sickly odor about her.

"Ruby, it's me, dear. Thank goodness, you came back." Charlotte bent over her dark-haired daughter, soothing her brow with a cool cloth that Victoria fetched for her.

Tears trickled from between her daughter's closed eyelids. Charlotte wiped them away with the cloth and kissed her forehead. "It's all right, dear. You'll be all right now."

Ruby opened her eyes and glared defiantly into her mother's eyes. "I won't either. Anyone can see I'm going to die. Even the doctor said so." She said sardonically, "I have a disease, mother. Lots of women like me get it."

Her eyes filled up with tears and ran over their red-rimmed edges. "I didn't want him to bring me here. I begged him not to bring me here to you."

"But why, Ruby? Why darling? Surely you didn't think I'd turn you away?"

Ruby spoke bitterly, "I didn't want to come back to you beaten and disgraced like a prodigal, expecting kindness for my sins. I won't blame you if you put me back out on the doorstep to die. I didn't love you, and you didn't love me either, so why pretend now?"

Tears trickled down Charlotte's cheeks and choked her voice. "Oh, my dear, how wrong you are! How very wrong! I always loved you. When you were a sweet baby with black ringlets and pink lips, and even when you left me. Yes, especially when you left me. I loved you so much, but there was poison in that house, and it consumed us all."

She spread Ruby's hair back on the pillow and put her cheek down against her daughter's. "And Ruby, darling, you may come back to me at any time, in any condition. You see, my dear, we are all prodigals, every one. The marvelous thing about love is that it takes prodigals and turns them into chosen sons and daughters. I don't care how you come to me, only that you come and let me love you. Will you, dear? Will you let me love you now at long last?"

But Ruby could not speak. Sobs were breaking through her weakened body, and she couldn't speak. But she circled her mother's neck with her arms, and Charlotte stayed with her throughout the night. She nursed her daughter faithfully for three more days, until Ruby passed back to the Father who accepts all prodigals. During those precious days, no bitter memories were mentioned, only the good. Charlotte recounted the times of wading in the stream together, picking berries, licking snowflakes from the darkened winter night. She recalled times only a mother could remember, when Ruby was just a baby at her breast. Ruby passed quietly, early in the morning before the dawn light, her hands clasped between her mother's.

The effort wore Charlotte down, and she spent the next months in bed herself, the consumption taking a greater and greater toll on her. Once a letter arrived from Annie and with it was a picture of her with her Aunt. Charlotte held it to her heart and cried for joy. The letter stayed by her bedside and the picture with it. Victoria did everything she could to stay the on-rushing disease, fixing all manner of herbal teas and special soups, and plastering her chest with odd poultices. But, despite all her efforts, Charlotte wasted away.

One day, in the late afternoon, there was a slight knock at the door. Victoria opened it and looked up into a thin haggard face, marked by dark hollow eyes and set off by a shock of dirty, black and gray hair. The man asked if Susanna Peterson lived here.

Victoria was suspicious and reluctant to answer, but finally admitted that she did, but allowed as how Susanna was ill. The man asked if he could see her briefly. He wouldn't tell her anything else but insisted on seeing her. He had such a force about him, Victoria could not withstand him. She let him in and then went to find Elberta. Together, she thought they could face him down.

Charlotte's door opened just a crack and into the shadows slipped Jack. She hardly recognized him, so bony and grisly had he become. He shuffled along the perimeters of the room, twisting a rotten excuse for a hat in his hand.

"So it is you," he whispered hoarsely.

"What do you want with me, Jack?" she asked quietly.

"Nothing. I only wanted to see if it was you. I heard about a woman, fitted your description, and I only wanted to see if it was you."

"Well, now you know." She watched him pacing nervously alongside the wall in the shadows. "I didn't think I'd ever see you again."

"Thought you'd killed me, huh? Should have finished it off, I guess. I'm just too mean to die easy."

He looked at her propped up in bed, pillows behind her back, and was uncomfortable in her presence. She had about her face a certain serenity and light. Maybe it was the afternoon sun throwing its rays on her face. At any rate, it made him uneasy. She was fifty years old, delicate—almost fragile— and her once-red hair was dull now with many silver threads. But in the hazy light, her face was still as lovely as the first day he had seen her swimming in the meadow pool. After a few minutes of silence he started toward the door.

"I'll be going."

"Why did you come, Jack? You hated me so."

He turned back to her, looked at her for a long moment, and then looked away. "Hated you?" His voice was gruff and hoarse, and he said painfully, "You're the only one I ever loved. And you never loved me. Nothing else has ever really mattered to me."

She watched him with pity. He didn't wait for her reply. He opened the door a crack and slipped out. A few hours later he was roaring drunk on the slum side of town.

One night, not long after, Charlotte had put down her reading and said her prayers. The only concern she now had was that she might one day, somewhere in the eternities, find that true love she believed in even yet. All her life, she had wanted to give love and have it returned. Only with her children had that come true. But Annie would belong to someone else, for that was the eternal way, a man and a woman sealed together forever. But she, Charlotte, had no one. Once or twice, she thought about that time long ago at Governor Hamilton's ball when she had thought she saw a man over by the column. She could remember clearly how he looked, gentle and loving, drawing her to him with a tender power, and she remembered a dream.

Again that night, she prayed she would one day be given to a man who would love her truly. She vowed she would bring him only happiness and a wealth of sweet, joyous love. She fell into a deep, peaceful sleep. About one o'clock in the morning, she seemed to see in her dreams a man watching her—a poet, she thought. It was the dark-haired, gentle man from the Governor's Ball. She roused from sleep and was surprised, mildly, to find him still there. Only, she couldn't see him clearly. He seemed to be waiting behind a veil, with his hand stretched out to her. A sense of joy swept over her, lifting her, transporting her, transcending all happiness she had ever known. Knowledge penetrated her spirit that this tender, loving man would be hers forever. She raised her arms to him happily. The man called her name and she went to him.

Victoria found her, peaceful and still the next morning. Her countenance spoke her final story, death for Charlotte O'Neill was sweet, very sweet!

THE END

THE LIGHTNING AND THE STORM is part one of a three part trilogy. In part two you will meet Charlotte's great granddaughter, Shielah, who is in love with two men— Stephen, tall, blonde and dedicated to the church, and Paulo, dark-haired, gentle, with the eyes of a poet or a saint. Is he the man of Charlotte's vision, or the love of Shielah's life? Finally, watch for part three where you will meet Charlotte again, and all their lives shall be intertwined in the Millenium.